CONDUCTING PRESCRIBED FIRES

Conducting Prescribed Fires

A COMPREHENSIVE MANUAL

John R. Weir

TEXAS A&M UNIVERSITY PRESS *College Station*

Second printing, 2014

This paper meets the requirements of
ANSI/NISO Z39.48-1992 (Permanence of Paper).
Binding materials have been chosen for durability.
♾

Library of Congress Cataloging-in-Publication Data

Weir, John Robert, 1964–
Conducting prescribed fires : a comprehensive manual /
John R. Weir. — 1st ed.
p. cm.

Includes bibliographical references and index.
ISBN-13: 978-1-60344-134-6 (pbk. : alk. paper)
ISBN-13: 978-1-60344-336-4 (ebook) 1. Prescribed
burning—United States—Handbooks, manuals, etc.
2. Burning of land—United States—Handbooks,
manuals, etc. I. Title.
SF85.6.F57W45 2009
634.9'55—dc22

2009004262

Contents

Preface vii
1 Why Conduct Prescribed Fires? 1
2 Law and Liability 13
3 Prescribed Fire Policy 22
4 Public Relations 27
5 Fire Weather 41
6 Fire Behavior and Fuel Characteristics 51
7 Fire Prescriptions 62
8 Fire Plans 71
9 Personal Safety 78
10 Firebreaks 88
11 Fire Equipment 106
12 Ignition Devices 123
13 Ignition Techniques 136
14 Smoke Management 156
15 Postburn Mop-up 168
16 Prescribed Fire Associations 177
References 183
Index 189

Preface

In the past twenty years I have conducted over seven hundred prescribed burns, and *Conducting Prescribed Fires* is the result of that experience. Many of these burns have been enjoyable, whereas others have been very educational and memorable, to say the least. It has been an honor to teach private landowners, college students, and agency personnel how to burn and a privilege to be associated with and learn from some of the best fire ecologists in the country, all of whom have greatly increased my knowledge and understanding.

I wrote this book to fill a gap in the current fire literature. Several books have been published on fire history, fire management, and fire ecology. But no book is totally devoted to an in-depth explanation of the requirements for conducting prescribed fires. A few fire books devote a section or chapter to conducting prescribed burns, but more information than can be found in a single chapter of a fire ecology book is necessary for someone to properly and safely conduct a prescribed burn. A prescribed burn entails more than just a fire plan and proper equipment. It incorporates law, policy, public relations, meteorology, safety, technique, smoke management, and numerous other elements.

As mechanical and chemical land management practices increase in cost or are determined to have negative environmental impacts, prescribed fire is viewed as an economical and viable method for ecological restoration and land management. Over the past decade the amount of federal funding has dramatically increased for wildland fuel reduction through the use of prescribed fire on public lands, cost-share programs for prescribed burning on private lands, and funding for fire research. With this increase in the use of prescribed fire and the public's greater awareness of fire throughout the country, it is important that the personnel implementing these fires have a comprehensive and up-to-date instructional guide.

I planned to write a practical, how-to book, accessible but technical enough to be used as a reference by natural resource professionals at all levels. All who read it, whether they have never conducted a prescribed burn or have been burning for years, should come away with newfound knowledge of and understanding for the use and implementation of prescribed fires.

I would like to thank my wife, Theresa, and children, Luke and Ashley, for their love, assistance, and understanding, and for tolerating the long hours, dirty clothes, and smell of smoke and diesel. I am also indebted to Dave Engle, Terry Bidwell, and Sam Fuhlendorf for their assistance, encouragement, oversight, and, most of all, their friendship. Finally, to everyone I have burned with over the years: thanks; I enjoyed it.

—J. R. Weir

CONDUCTING PRESCRIBED FIRES

CHAPTER 1 Why Conduct Prescribed Fires?

But, that which beareth thorns and briers is rejected, and is nigh unto cursing; whose end is to be burned.

—Hebrews 6:8

Historically there have been two main causes of fire in the world: lightning and humans. Lightning has been a source of ignition in a majority of the ecosystems around the world; however, most lightning fires do not burn large areas or turn into conflagrations as anthropogenic, or human-caused, fires do. So the effect of lightning fires may be small for an entire continent, but the localized effect is still significant (Pyne 1982).

Anthropogenic fire has been shaping the world's inhabited ecosystems since the beginning of time; the first recorded use of fire in the Bible occurs in the third chapter of Genesis. Historical aboriginal fire involvement spurred the development of a humanized landscape, even though most people envision the past as a pristine or non-human-made landscape (Vale 2002). Many researchers state that no pristine wilderness existed in the United States when European settlers first arrived; instead, the landscape consisted of prairies and forests created by indigenous peoples (Martinez 1988) who exploited and manipulated the landscape with fire (Lewis and Ferguson 1999). In much the same way we view the land today, some Native Americans considered lands not managed by humans as wilderness (Anderson 2005). Native Americans used fire for numerous reasons in their daily lives; in fact, evidence shows that they burned the land for more than seventy different reasons (Lewis 1973).

The Native Americans' use of fire explains the early explorers' comments of vast expanses of tallgrass prairie and detailed descriptions of pine or hardwood savannas throughout the eastern and western United States. These plant communities are all fire-driven ecosystems that are vanishing from their original ranges due to decades, and even centuries, of fire suppression.

Changes in the use of fire came as Europeans settled North America. When Europeans immigrated to the United States, many were fascinated by the use of fire in Native American culture, but in most regions fire was still feared and suppressed as it had been for centuries in Europe. Settlers in the southeastern United States continued to use fire in much the same way that the Native Americans had for the first four hundred years after their arrival (Williams 1992). Before the Civil War over 75% of the Caucasian population in the southeastern United States consisted of pastoral herders of Celtic origin from England, France, and Spain, where fire had been an important part of their nomadic culture (Owsley 1945; McWhiney 1988). This fire culture still remains throughout most of the Southeast today but is lacking in many other U.S. regions.

After the beginning of the twentieth century a stronger demand for fire suppression emerged throughout the United States. During World War II fire suppression was put on a national priority list for the defense of the nation. However, during this time large wildfires occurring in the South caused people to realize that prescribed fire was the only way to suppress the larger threat of unintended fires. In 1943, following the aftermath of severe fires from the previous year, the first approved prescribed fire on a southern forest was conducted on the Osceola National Forest in Florida (Stanturf et al. 2002), which started the slow process of introducing fire back into ecosystems and to a public with an innate fear of fire.

Prescribed burning has not been completely accepted and still has many opponents to this day. When conducting prescribed fires today, we need to remember the sacrifices that early fire professionals made to give us the knowledge and ability to properly conduct burns. Prescribed fire should be used in a responsible manner, and nothing should be done to damage the image of prescribed burning that has taken decades to develop.

Why Burn?

One of the best reasons to use prescribed fire is the usually lower cost of burning compared to other land management options, such as mechanical or chemical treatments (Bidwell, Weir, and Engle 2002). When lands have undergone years of mismanagement, prescribed fire alone is not the most effective management method. In such cases, you as land manager may have to incorporate mechanical or chemical treatments along with prescribed fire to reclaim certain areas (Figure 1-1). Combining treatment options allows prescribed fires to be more effective. Ultimately, the land manager must choose which treatment is best suited to the goals and objectives on a particular parcel of land. Numerous studies throughout the United States have compared different land management techniques and their ecological impacts. To find out which treatments would be the most effective and economical in your area, consult your local cooperative extension service, Natural Resource Conservation Service (NRCS), or land management consultant for specific information.

LIVESTOCK PRODUCTION

One of the main reasons many people use prescribed fire is for the enhancement of livestock habitat. When you create better habitat, you can usually increase individual animal performance and possibly increase carrying capacity. One of the main ways to accomplish this is through the control of woody plant species that are invading rangelands throughout the United States.

A significant woody plant problem occurring throughout much of the Great Plains is the invasion of eastern redcedar (*Juniperus virginiana*). This juniper species primarily causes a drastic decline in forage production. This reduction in available forage, in turn, affects carrying capacity, stocking rate, and

Figure 1-1.
Once vegetation is in the reclamation phase, prescribed fire becomes less effective. You may be required to use prescribed fire along with other methods, such as this ball and chain, to reclaim the landscape.

livestock performance. In fact, a range site with a potential to produce 4,000 pounds per acre (4,480 kg/ha) of forage, with a population of 200 eastern redcedar trees per acre (494 per ha), would produce only 3,700 pounds of forage per acre (4,144 kg/ha). If that site is left unmanaged, in ten years the population could increase to 470 eastern redcedar trees per acre (1,161 per ha), and the site would produce less than 2,000 pounds per acre (2,240 kg/ha) of forage (Engle and Stritzke 1992b). With the reintroduction of prescribed fire at two- to five-year intervals, this problem could be removed and the land could return to its production potential.

Prescribed burning does have limitations for woody plant control, especially if you wait too long to reintroduce fire. Many plant species will gradually grow too large to be impacted by typical prescribed burns. Eastern redcedar is a prime example. Prescribed fire has been shown to be the most efficient and economical method for controlling this invasive, nonsprouting native juniper (Bidwell, Weir, and Engle 2002), and nearly any type of fire will kill trees that are less than 1 foot tall (0.3 m). Note that fine fuel loading is one of the most important considerations of any burn when the objective is to control woody vegetation. Greater fuel loads will allow the burn to be conducted under safer conditions (lower temperature and higher relative humidity) and still achieve your goals. Eastern redcedar trees from 1 to 5 feet tall (0.3–1.5 m) can be controlled very well when herbaceous or fine fuel loads are above 4,000 pounds per acre (4,480 kg/ha), but if the fuel load is reduced by half, only 60% of the trees less than 5 feet tall (1.5 m) are killed (Bidwell and Stritzke 1989). Then, as the eastern redcedar trees grow taller than 5 feet (1.5 m), they are even harder to control, requiring a greater fine fuel load or a more extreme burn prescription. Using an extreme prescription would mean burning under higher temperatures and lower relative humidity, and a greater risk for an escaped fire. Therefore, maintaining adequate fuel loads through proper stocking rates is the most important part of any prescribed fire and grazing program.

Other brush species are also overtaking or invading U.S. grasslands, shrublands, and forestlands. Most of these species resprout after a fire, including oaks (*Quercus* spp.), sumac (*Rhus* spp.), mesquite (*Prosopis* spp.), dogwood (*Cornus* spp.), and other junipers (*Juniperus* spp.). Although these species are native, they are increasing in height, density, and coverage area. Fire typically reduces the height of these plants but increases the number of stems due to their propensity to resprout. Repeated fires will keep most of these plants suppressed or even remove certain species from the system altogether.

A commonality in fire frequency has been found when comparing sand shinnery oak (*Q. havardii*) in western Oklahoma grasslands to oak–shortleaf pine (*Pinus echinata*) forests in southeastern Oklahoma and pine forests of northern Florida. Even though these sites are hundreds of miles apart, and precipitation ranges from 20 inches (508 mm) at the western site to over 50 inches (1,270 mm) at the southeastern site in Oklahoma to over 60 inches (1,524 mm) at the Florida sites, these studies concluded that a fire return interval of three years or less should be used to manage woody plants and keep them suppressed (Masters and Waymire 2000; Boyd and Bidwell 2001; Masters et al. 2005) (Figure 1-2).

An important point to remember is that although prescribed fire will not solve all of your land management problems, it can solve some of them. Also keep in mind that prescribed fire is not a one-time treatment but a program that must be continued, and fire frequency is the key to success.

Another major benefit of using prescribed burning for livestock production is the increase in weight gain of stocker cattle. Ranchers in the Osage Hills of Oklahoma and the Flint Hills of Kansas have used fire to this end for years. Studies have shown that prescribed burning increases summer stocker cattle gains by 10% to 20% in the summer following the burn (Anderson, Smith, and Owensby 1970; Smith and Owensby 1972; Wolfolk et al. 1975; Owensby and Smith 1979). Researchers have published little work on the effects of prescribed fire on cow/calf operations. People who conduct research on cow/calf operations and use fire to manage their land have noted that cows can usually

Figure 1-2. These photographs illustrate how important fire frequency is for the control of shrubs and trees. The photos on the left are from southeastern Oklahoma in oak-pine forest (photos courtesy of Ron Masters); those on the right are from sand shinnery oak grassland in western Oklahoma. Both sets depict an area with no fire (top) and a two-year fire frequency (bottom).

increase their body condition score by a factor of 1 over cows on unburned range; researchers have also noticed calves with increased weaning weights as a possible side benefit.

There are times when brush or trees should be left growing in an area. Historically fires kept some rangeland areas as savanna or parklike settings. These lands have now become closed canopy shrublands or forests due to years of fire exclusion. This has allowed brush to invade the interspaces or openings, thus causing a reduction in forage and livestock production. Reintroducing fire into such a system can open it up, once again making it a suitable habitat for livestock.

Burning will cause an earlier green-up of plants following a dormant-season burn because the blackened soil surface resulting from the fire's removal of old forage growth allows soil temperatures to warm up earlier. Fire can also be used to remove old forage growth to improve the palatability and increase the nutrient content of new growth, as well as an overall increase in the diversity of the plant community.

When livestock will not frequent certain areas within a given pasture, fire can help with the distribution of livestock. This aversion to certain areas may be due to old growth of grass, previous grazing patterns, topography, distance from salt or water, or type and age of livestock traditionally used. If fire is used to improve the quality and availability of the forage, then livestock should begin to use these areas again. Animal performance can improve, and carrying capacity may increase; forage quality and quantity of the area may also improve.

Prescribed burning will generally assist with overall land management and facilitate the best use of the land by livestock. In areas without fire, it is normally difficult to observe and oversee the health of livestock because of shrub or tree cover. One rancher who uses prescribed fire noted that it was worth the cost of burning just to be able to drive into a pasture and see all of his cattle. These fire effects can also be very helpful when you are gathering livestock. Many times livestock that live in brushy areas will use the dense cover to evade capture. But as fire opens up a pasture, the livestock will have fewer places to hide. Whereas past gatherings may have required a half-dozen cowboys and twice as many dogs, now you need only a pickup truck and feed sack because the cattle can be seen more often.

Prescribed fire can also be used to control external parasites, such as ticks, that affect livestock. For tick control to be effective, the heat during the burn must reach an intensity and duration high enough to both kill the tick population and remove the cover (litter) that protects surviving ticks (Buck 1935; Heady 1960; Van Amburg, Swaby, and Pemble 1981). (The intensity and duration are discussed further in Chapter 7.) The size of the burn unit (the area to be burned) will determine how rapidly these pests are brought back into the area by other animals after a prescribed burn. Burning small areas will not control ticks as effectively as burning larger areas because it takes longer for the area to become repopulated after a larger burn. Fire has also been shown to reduce certain species of internal parasites that affect livestock by interrupting their life cycle outside the host.

Wildlife Habitat

Prescribed fire is a very important management tool for wildlife habitat improvement, reclamation, and maintenance of vegetative structure required by fire-adapted species. When using prescribed fire to manage for wildlife, be sure to consider the needs of all species that utilize the habitat. Volumes of research have been conducted on habitat requirements and fire effects for specific wildlife species. You can access the specific requirements for the wildlife species you are trying to manage by contacting your state wildlife department, NRCS, county cooperative extension service, or private wildlife consultant.

As stated previously, control of invasive brush species, such as eastern redcedar, is important for maintaining the proper environment for certain wildlife species that can live only in specific habitats or are associated with a specific plant structure. These species are known as habitat specialists. Many habitat specialists, such as grassland obligate bird species, cannot coexist with an increase in the coverage or height of trees and shrubs in their landscape. Certain prairie songbirds will leave an area when the eastern redcedar cover reaches as low as 10% of trees per acre (Chapman et al. 2004). Research also shows that bird species' richness will increase in native grassland areas as woody plants increase; this species increase is due to grassland species being replaced by shrub and forest bird species (Knopf 1986; Coppedge et al. 2001). This research shows why brush control is so important in habitat and specific species' management planning.

Burning will increase the amount of bare ground on a given site, which is important for many upland birds and small mammals that rely on seeds for food. The seeds must be visible in order for the birds and mammals to utilize them, and removing the mulch layers makes the seeds readily available to these species. Clearing the ground also enables some species of small mammals and ground-dwelling birds to move about with greater ease. On the other hand, some animals require the thick mulch layers to survive; remember to keep these species in mind when burning and avoid burning the entire area. Be sure you understand the habitat requirements for specific animals and the possible impacts to other species in the area when you conduct prescribed fires.

Forbs, which are broadleaf plants, are an important component of many wildlife species' diets. Burning at different times of the year can increase the amount of forbs or can cause different species of broadleaf plants to emerge. Keep in mind that burning repeatedly during the same time of year will promote certain forbs, but if you change the time of year that you burn, you can encourage many other species to grow that normally would not occur or would not exist in large enough numbers to make a difference.

Fire can be used along with grazing to create a mosaic or heterogeneous landscape that is important to some animals (Fuhlendorf and Engle 2001). Some species require areas that are ungrazed and unburned, while others may require areas that are burned frequently and grazed heavily. Some species, such as the Lesser Prairie-Chicken (*Typanuchus pallidicinctus*), require a broad range of habitats for survival. For nesting, the Lesser Prairie-Chicken requires areas that have not been burned or grazed for two years; for brood rearing, it prefers areas that have large amounts of forbs. These weedy areas are created in the year following a fire and after heavy grazing has occurred. The Lesser Prairie-Chicken also requires areas with very little vegetation, called booming or leking grounds, for their breeding displays. These booming grounds are found in areas that have been burned and grazed heavily. Finally, the Lesser Prairie-Chicken does not like vertical obstructions such as trees or tall shrubs anywhere within its home range. Frequent fires can remove such obstructions or reduce their stature (Bidwell, Fuhlendorf, et al. 2003).

Prescribed fire should be used to maintain native habitats for native wildlife species. Most all U.S. habitat types originate from fire-derived ecosystems, but fire has been removed from many of these habitats for decades or even centuries. Fire can also be used to remove many invasive or alien plants that can degrade these native wildlife habitats. Remember that some alien plant species will increase with burning, so burning at different times of the year may have a greater impact on these plant species. You may also have to incorporate other control methods, such as herbicides, into your fire management program to suppress invasive plants.

Browse from trees and shrubs is very important to many wildlife species. Prescribed fire will cause most browse species to resprout once the fire has top-killed the old growth. Fire allows these plants to produce tender new growth that is within reach of animals. The benefits of prescribed fire with regard to browse are substantial for wildlife in the first several years following a burn; fire reduces stand density and makes browse more accessible. Fire also reduces the height of the browse, allowing more of it to be grazed by animals (Canon, Urness, and DeByle 1987; DeByle, Urness, and Blank 1989). Fire has even been shown to increase mast production of some oak species. Because acorns are very high in nutrients, they are important to many wildlife species. Research in southeastern Oklahoma has found that a four-year fire frequency along with stand thinning increases mast production on post oak (*Q. stellata*) and blackjack oak (*Q. marlandica*) (Masters and Waymire 2000).

Fire can also be a useful tool for controlling parasites that affect wildlife. As mentioned previously with livestock production, fire can be used to control ticks and internal parasites. Fire has been found to be an important control method for one parasite that can be fatal to certain wildlife species, the meningeal brain worm (*Parelaphostrongylus tenuis*). This worm has caused numerous deaths in elk (*Cervus elaphas*) that have been reintroduced to the eastern United States (Carpenter, Jordan, and Morrison 1973). The parasite lives in certain terrestrial slugs and snails (gastropods) that elk unintentionally ingest as they forage. These gastropods are dependent upon moist forests with high tree densities and a well-developed midstory (Raskevitz, Kocan, and Shaw 1991). The meningeal brain worm enters the elk's stomach, where it then migrates to the brain and spinal cord, damaging these central nervous system tissues and ultimately causing death to the host (Fraser et al. 1991). Fire opens the forest up and dries out the microclimate, thus reducing the gastropods' habitat. Another added benefit is that more open and drier habitat is better suited to, and preferred by, elk.

Snags for cavity-nesting birds or animals that make dens in trees can be created through the use of fire. However, these snags can also be burned up during prescribed fires. Exercise care when burning around snags so they do not catch fire. Rake around and backfire away from the snags, or wet the snags down with foam or water prior to ignition. The main thing to remember is not to burn all the cover in a large area. Loss of cover can impact nests, as well as young or slow-moving animals. Be sure to plan your burns carefully to reduce the fire's impact on any of the life cycles of wildlife in your burn area.

Although you will want to be aware of the needs of the species living in your burn area, do not be overly concerned about trapping too many species within your fire because most can easily evade the flames. Infrequently, a fire-induced mortality may occur on a single animal, but it is better to sacrifice one animal for the benefit of the entire population than to worry about the loss of a single individual. You have to keep in mind what will eventually happen to the health of the habitat and wildlife population overall if you do not burn. Most of the mortality to wildlife that I have observed has been on the herptofauna. Snakes, lizards, and tortoises are slow moving and seem to have higher fire mortality rates than other wildlife species, but most of these deaths can be avoided by burning during cooler months when these animals are hibernating. Therefore, if the management or conservation of an individual species is your main objective, plan your burns when there will be less impact on that population.

Forestry

Fire has a large application for forest management throughout the United States. Prescribed fires can be used to thin dense stands of trees, eliminate young trees, and maintain park or savanna-like openings (Wright and Bailey 1982). Prescribed burns will thin stands, thus allowing the remaining trees to increase in diameter while eliminating thicket areas (Weaver 1967). Prescribed fires also help reduce wildfire potential, and in certain areas this protection can last for five to seven years following the burn (Truesdell 1969). Wildfires that do start in areas managed with prescribed fires are easier to control and cause less damage to standing timber than do wildfires in unmanaged areas. In New Mexico and Arizona, damage to trees by wildfires was significantly reduced by previous prescribed burns (Cram et al. 2008).

Fires will promote growth of fire-tolerant tree species native to the area. Many tree species are fire tolerant because they have reproductive mechanisms such as underground rhizomes, root sprouting, and serotinous cones that help them survive periodic fires. Some fire-tolerant trees have thick bark and epicormic sprouting to assist them in surviving fires. Fire will also help promote the growth of other forest plants adapted to fire. With the suppression of fire in many areas, numerous native plant species have declined or disappeared. The reintroduction of fire has brought many of these species back. One disadvantage of frequent prescribed fires can be that they are usually conducted during the same season every year. This type of fire regime can promote certain species and be detrimental to others. It is important to rotate your burning season every so often to promote the growth of different plants. Depending upon the intensity and frequency of fires, prescribed burning is beneficial for forest health because it encourages natural regeneration. This is accomplished by seedbed preparation, as well as scarification of serotinous cones and other seeds. Fire can also be used for hardwood reduction in pine forests, which will reduce competition between pine seedlings and hardwoods. Prescribed fires will reduce competition for water, space, sunlight, and nutrients by thinning out the plant species and will help recycle nutrients, making them available for the new growth. Some species of trees are very susceptible to high-intensity fires, so be sure to burn with the proper technique to reduce crown scorch.

Fires can be conducted to control many forest disease problems (Froelich, Hodges, and Sackett 1978; Wade and Lunsford 1989; Thies 1990). Many times the fire destroys the infected tree or tree part, or the fire can change the microclimate on the forest floor, which may destroy the disease. Fire can also be used to control some forest insect problems and has been shown to be more cost effective than chemical controls for certain insects (Simmons et al. 1977; Miller 1978; Wade and Lunsford 1989; Mitchell 1990).

Although management of logging areas may not be the first application people consider in relation to prescribed fires, prescribed burning can be used before a logging operation begins to allow crews easier and more efficient access to an area when cruising and marking timber. The cleared areas make harvest operations more efficient and increase the equipment operators' visibility and safety by allowing them to see hazards more easily (Wade and Lunsford 1989).

After a logging operation is finished, limbs, undesirable trees, and plant stems are left on-site. This debris causes extreme difficulty for land managers accessing these areas and makes it difficult for planting equipment to maneuver. If a wildfire were to occur in such an area, the logging debris would make fire control very difficult and fireline construction impractical. Clearing the area increases forest access for recreational users.

As an added bonus, burning will open up the understory of many forest settings and thus allow livestock production. Livestock grazing adds value to the property and will lessen fuel loads, causing a reduction in wildfire potential.

Wildland/Urban Interface

Prescribed fires can be used to protect areas within the wildland/urban interface zone. They have been used to remove fuels in areas with large quantities of biomass. These fires are normally conducted just prior to the critical wildfire season to reduce the amount of fuel, making fire control easier and safer for firefighters. In Los Angeles County, California, firefighters have developed a technique that uses both crushing and burning of the fuels. A bulldozer is driven over the chaparral to crush it; the vegetation is then burned four to eight weeks later. The crushing process reduces flame lengths to 2 to 3 feet (0.6 to 0.9 m), instead of the 60-foot (18.3 m) flame lengths common in the standing chaparral. This technique also produces lower smoke emissions (Franklin 1993). In Florida, research has shown that prescribed burning in palmetto-gallberry fuels reduces the number of acres burned by wildfires and that the number and average size of wildfires are diminished (Martin 1988; Koehler 1992). Along with benefits of protection of property, a Louisiana community has received a savings of 5% to 10% on insurance premiums because it uses prescribed fire for wildfire prevention (MacKenzie and Fortier 2004).

Prescribed burning can be use to convert volatile fuels to nonvolatile fuels or remove the volatile fuels altogether. This process is used throughout the southern Great Plains in areas heavily infested with eastern redcedar. During wildfire episodes, dry eastern redcedar is an extremely flammable fuel containing aromatic oils and phenolics. When these volatile trees ignite, or crown, they can send burning embers great distances downwind, causing spot fires that make control efforts extremely difficult and unsafe for all personnel involved. Also, when the trees are crowning, firefighters cannot get close enough to extinguish them. The use of periodic prescribed fires, or in some instances mechanical treatment plus fire, will remove this volatile fuel problem from most areas. This problem is not limited to the southern Great Plains but can be found everywhere there is an increase in flammable brush species.

Fire can also be used to change the structure of the fuels in an area. Areas that have large amounts of tall fuels and dense understory or ladder fuels can be altered with low-intensity, periodic prescribed fires. Removing these fuels will reduce the threat of crown fires and spot fires, thus making wildfire suppression safer and easier (Figure 1-3).

Using prescribed burns to create safety or buffer zones around homes, housing additions, or towns is a technique that has been used for centuries. This blackened or green buffer strip provides a no-burn zone to protect property and lives. It also gives firefighters a safety zone from which to fight a fire, as well as a travel corridor for equipment.

There are numerous reasons why prescribed fire should be justified as the appropriate landscape management tool, one of the most important being liability. If you were asked in a court of law why you were setting fires, could you explain your actions in a way that a jury could understand? Being able to show that you put some thought into your land management choice before you began burning can go a long way toward convincing a jury that you tried to mitigate your risks while burning. It is very important to have documentation explaining why the prescribed burn is being conducted. The documentation can help show that your actions were well thought out and based on scientific fact. Many times concerns about liability issues dictate how, when, and why prescribed fires are either implemented or not.

Aside from liability issues, you should be concerned about satisfying public concern when you justify using prescribed fire. Whether we want to admit it or not, what other people think about the activities we are involved in is important and many times dictates which land management practices we use. Public opinion can change how private land managers or government agencies manage their resources. If a community has not been exposed to prescribed

Figure 1-3.
Fuel reduction, along with changing the type of fuel, is good motivation to use prescribed fire in the wildland/urban interface. The light area in the center is a house.

fire before, or if the community is afraid of using fire to manage the land, you need to be ready to educate the public on the positive reasons for prescribed burning. You will probably need to answer questions and help assuage negative public opinion with realistic reasons for using prescribed burning.

Whether you work for a public agency or private organization, your supervisor will want to have a good reason for burning. If you cannot demonstrate the positive benefits of using prescribed fire, then you will not be able to burn. Also, if something negative or unexpected should happen on your burn, the support of your superiors is very important. Their support could mean the difference between being personally liable for damages or losing your job, or quite possibly both.

Whatever the reason you decide to burn, make sure you conduct the fire under safe conditions with the proper equipment and adequate personnel. Fire is an effective and economical technique for natural resource managers to use. Fire is the only naturally occurring management tool we have. In fact, you could say that prescribed fire is completely natural and environmentally friendly.

CHAPTER 2 Law and Liability

If a fire break out, and catch in thorns, so that the stacks of corn, or the standing corn, or the field, be consumed therewith; he that kindled the fire shall surely make restitution.
—Exodus 22:6

Written fire law has been around since the time of Moses, and Exodus 22:6 states that the liability from an escaped fire rests on the person setting the fire. Even today, according to landowners in several states, the number-one reason they do not burn is fear of liability. A lot of the public's fear is unfounded because of their misunderstanding of prescribed fire. Most people associate all fire with wildfires; they think that if a fire escapes, it will burn uncontrollably and consume the entire countryside. Many people believe that all fires behave the same way large wildfires behave.

After talking with people who have given expert testimony for numerous court cases involving fires in Oklahoma, I have not found any lawsuits involving a properly conducted prescribed fire that resulted in a large settlement. An example of a typical lawsuit is the following 1999 case tried in Osage County, Oklahoma (*Lowe vs. Jones et al.* Case No. CJ95-345), where the plaintiff sued for $9.3 million in damages on a 200-acre (80.94 ha) fire that was claimed to have resulted from a prescribed fire (Yoder 2002). The fire burned only grassland, no structures. There was no judgment found for the plaintiff, but fees for the defense reached close to $0.5 million. So, the fear of large lawsuits may be unfounded, but the cost of defending yourself in court is still a possibility even if you conduct your burns properly (Figure 2-1).

You must also remember that some people are greedy and will usually only pursue litigation if the defendant is a person or company with substantial financial backing. Such was the case when a prescribed fire contractor conducted a burn for a private landholder. The holder of a livestock-grazing lease hired the private contractor to burn a specified parcel of land. A different person also leased the same land for the hunting rights. The hunter had an old, dilapidated cabin on the land, and the livestock lessee told the hunter to clean up around the cabin in preparation for the prescribed burn. During the prescribed fire, the old cabin ignited and burned down because nothing had been done to protect it. The hunter called the private contractor to see what the contractor was going to do about the cabin. The contractor told the hunter that it was between him and the livestock lessee, since the livestock lessee signed a waiver assuming liability for the burn.

The hunter asked if the state university had conducted the burn (the university owned land adjacent to this property and conducts numerous prescribed burns) and was informed that the burn was not conducted by nor in any way connected to the university. The hunter then told the contractor he had hoped

Figure 2-1.
Is our fear of liability founded on fear itself? How many large lawsuits have actually occurred because of damage from properly conducted prescribed fires?

the burn was the university's, because he thought the university would pay several thousand dollars for the cabin. When the hunter found out that the university was not involved, he dropped all inquires about gaining restitution. This story shows that if you do not have extensive financial resources, or if you are an individual and not a large company, you are not as likely to have someone sue you. Bringing suit against a corporation or agency is less personal, which may explain why people are more willing to sue corporations and agencies than individuals in their community.

Liability Laws

Three types of liability exist in prescribed fire laws used in the United States. Some laws are favorable to prescribed burning, while others are written to limit the use of fire. Written laws may be black and white in their statement of liability, but a jury ultimately determines negligence and the amount of damages awarded. Even in lawsuits involving wildfire against large companies, greed loses to science and fact more often than not. These are the three types of fire law:

- *Strict or unlimited liability:* An activity that is not so unreasonable as to be prohibited altogether, but the activity is sufficiently dangerous or provides unusual risks. Therefore, the law requires the activities to be conducted at the peril of the person or group sponsoring the activity. In this situation, the person setting the fire (defendant) will be liable and have to pay for damages even when there is no evidence of negligence on the defendant's part. The victim (plaintiff) will only have to prove that monetary damages occurred and that the defendant's actions (prescribed fire) caused the damages. At this time four states have this type of law: Oklahoma, North Dakota, New Hampshire, and Connecticut (Yoder et al. 2003).
- *Negligent unless proven otherwise:* Burden of proof is on the prescribed burner (defendant) if the fire escaped and caused damage to someone else's property. In other words, you are guilty until you prove yourself innocent. Twenty-two states have this type of law, and five of them have laws that cite an escaped fire as prima facie (or legally sufficient) evidence of negligence (Yoder 2002) (Table 2-1). To escape liability, the prescribed burner must prove that due care was exercised while conducting the burn, because sixteen states place the burden of proof for negligence on the defendant (Yoder et al. 2003). Oregon is in both categories: its law states that the prescribed burner is liable for double the amount of damages if negligence is proven and single damages if there is no proof of negligence (Yoder et al. 2003).
- *Not negligent unless proven negligent:* The burden of proving the prescribed burner (defendant) negligent rests on the victim (plaintiff). The burner is considered innocent unless proven negligent. Florida, in 1990, was the first state to pass this type of law (Brenner and Wade 1992). At this time six other states have passed similar laws: Alabama, Georgia, Louisiana, Mississippi, South Carolina, and Texas (Table 2-1).

All of these laws have their strengths and weaknesses. In a "strict or unlimited liability" state like Oklahoma a person can burn anytime, at his or her own peril, with no stipulations from authorities. In a "negligent unless proven otherwise" state such as Nebraska, conducting prescribed burns requires a permit issued by the local fire marshal, who many times has no understanding of prescribed fire and in some areas is not very willing to grant permission. In Florida, a "not negligent unless proven negligent" state, prescribed burners must attend a week-long training class, conduct a certain number of burns, and receive authorization from the state Division of Forestry before becoming exempt from liability. For your own protection, be sure to read and understand the laws within the state where you are conducting prescribed fires.

The main downfall of "strict or unlimited liability" is that it leads to less risk mitigation by adjacent landowners than an efficient negligence rule does

TABLE 2-1 State liability laws for prescribed fire and fire risk

Liability or property rule	*State*
Burner strictly liable	CT, ND, NH, OK
Burner presumed negligent if fire escapes	AK, GA, MD, OR, UT
Burner liable for damage if proven negligent	AL, AR, CA, DE, FL, LA, ME, MI, MS, NC, NJ, OR, TX, VA, WA, WI
Prescribed fire act or law	AL, FL, GA, MS, NC, SC, TX, VA
Permits or bans supported by statute	AL, AZ, CA, CO, CT, FL, GA, IA, ID, MA, ME, MN, MS, NE, NH, NJ, NV, NY, OR, RI, SD, UT, VT, WV, WA
Criminal penalties for leaving fire unattended or failure to extinguish and negligent escape	CA, MI, NC, NJ, NM, NV, OK, OR, SC, SD, TN, UT, WI, WY
No statutes addressing prescribed fire	HI, IL, IN, MO, MT
Liable for negligently allowing uncontrolled spread of wildfire	AK, DE, MI, OH, OR, PA, SD, TN, TX, UT, VT, WA, WV
Uncontrolled fire is a nuisance: can be billed for public fire suppression costs	CO, GA, ID, MD, ME, MS, NH, OK, OR, WA, WI
Regulations restricting excessive vegetation fuel loads	MN, MT, NM, WA

Source: From Yoder et al. (2003).

(Yoder 2002) (Figure 2-2). Thus, in an area with strict liability, neighbors are going to take very little precaution in protecting their property from fire, because they know someone else will pay for the damages. This is different from a "not negligent unless proven negligent" state where homeowners and landowners know they are responsible for protecting and recovering their property from escaped fire unless they can prove negligence.

Some people believe that the "not negligent unless proven negligent" laws of the southern states may yet be challenged and reshaped in the courtroom, but one burner in Georgia has already been successfully defended in court under the Georgia Prescribed Burning Act (Brenner and Wade 2003). Throughout the United States, several civil cases have headed to court but were dropped or settled by the plaintiffs at the last minute (Brenner and Wade 2003). Therefore, these laws are protecting those who properly use prescribed fire.

Figure 2-2.
"Strict liability" prescribed fire laws lead to less risk mitigation by adjacent landowners, as illustrated by these homes built in the midst of volatile fuels with no apparent defensible space around them. This is also a prime example of contributory negligence on the part of the homeowner (photo courtesy of Terry Bidwell).

An example of a weakness in a "not negligent unless proven negligent" law occurs in Texas, which has a prescribed burning act and certification for prescribed burners, but its law is being held in limbo by insurance requirements. The Texas law requires that all certified prescribed burners have a $1 million liability policy. At this time, no underwriter will sign for this insurance, so there are no certified prescribed burners in the state. This is a case of a good law being handicapped by unobtainable requirements.

One type of law that would most benefit prescribed burning is one addressing contributory negligence, which results from people contributing to their own harm. Two states, Connecticut and Illinois, currently have a contributory negligence statute addressing fire (Yoder et al. 2003). An example of contributory negligence would be allowing volatile fuels to grow in close proximity to a home or outbuildings, or using combustible building materials in home construction (i.e., wood shingles, wood siding) and then sustaining damages from a fire (prescribed or wildfire). A contributory negligence law would assist prescribed burners by making people aware that they are responsible for paying for part of the damage to their home, buildings, or lands if a fire were to escape.

At this time, one major insurance company has initiated a contributory negligence clause in their homeowner policies in Arizona, Nevada, and New Mexico. Insured persons in the fire-prone areas of these states must comply with rules defining proper fireproofing around their homes. The insurance company inspects the insured homes and makes recommendations to the homeowners. The homeowners then have a specified amount of time to comply with the recommendations, at which time the insurance company will then reinspect the homes. Insurance companies will cancel policies of homeowners not in compliance with the recommendations.

Managing Liability

Managing your liability is ultimately up to you, the person conducting the prescribed fire. The four ways to manage your risk are as follows:

- *Elimination:* Avoidance and discontinuance are the two forms of elimination. Avoidance identifies the risk (prescribed fire) of a situation occurring (escape). After assessing the risk, someone decides not to use prescribed fires to manage the land at that time. (Avoidance is very popular among many landowners, supervisors, and administrators.) Discontinuance occurs when the activity (prescribed fire) is said to be too great a risk and is no longer offered as a land management option. This choice has allowed our landscape to reach the current state in which catastrophic wildfires are increasingly prevalent.
- *Transfer of risk:* The financial risk is transferred to another person or entity by contract or affiliation agreement (van der Smissen 1990). One way to transfer risk is by contracting out the prescribed burning work to a third-party contractor; the contractor may retain much of the liability and may also carry some type of insurance (Stanton 1995). The first problem with transference of risk is finding a contractor who conducts prescribed burns in your area. Second is finding an insured contractor, as evidenced previously by the reference to Texas law. At present, finding and affording insurance are major obstacles for private prescribed burning contractors. Third, the person or entity hiring the contractor can still be held liable for the contractor's negligence. The employer should exercise reasonable care to choose a competent, experienced, and careful contractor who uses proper equipment and takes appropriate safety precautions.

The other way to transfer risk is to carry insurance to cover litigation that may arise from escaped fires. You must remember three things about insurance: payments on the premiums must be up-to-date; insurance provides only a certain amount of coverage; and having insurance does not prevent lawsuits but only provides some financial coverage in case you are sued (van der Smissen 1990). Sometimes insurance may even encourage lawsuits.

Most farm and ranch liability policies for private landholders cover damages from escaped prescribed fires. Check with your insurance provider for your specific coverage information. You may also want to increase the amount of your liability coverage. Often, you can increase coverage from $0.5 million to $1 million with only a slight increase in your premiums; again, check with your insurance provider for specifics. Insurance coverage is usually not an option for state or federal agencies. Most of these groups are self-insured, and insurance cannot be purchased. Some states also have a limit on the amount of restitution for which their agencies are liable.

- *Risk retention:* The burner assumes responsibility for the potential financial loss from an escaped fire. Retention means including the risk as part of your own financial planning. Governmental agencies use this type of risk management. A majority of the risks are expected, but some are unplanned, and the agency tries to reduce the magnitude and frequency of those losses. Usually an agency establishes a funded reserve when it wants to be in charge of its own financial risks, sometimes referred to as being "self-insured" or having a contingency fund. In the long run, this reserve can result in a financial savings to the agency.
- *Risk reduction or risk management:* A landholder, organization, or agency will determine that using prescribed fire is hazardous and set up policies and training to reduce the probability of accidents (escapes). Most federal and state agencies practice risk reduction in all of their hazardous activities. An example is an agency recognizing the fact that prescribed fire can cause damages or personal injury and requiring all persons involved with prescribed fires to be trained and/or certified according to a certain set of standards. The agency will also have a written policy dictating when, how, and under what conditions a prescribed fire may be set, as well as what equipment is essential.

If you are a private landholder, you can also practice risk management in several ways. You can receive training by attending workshops or classes hosted by universities, state agencies, or organizations. You can also develop your own policy or guidelines under which you will or will not burn and have a written fire plan for all burns. Either of these options can help demonstrate your commitment to safe burning and possibly prove in a court of law that you were not negligent. Probably the best way a private landholder can conduct risk management is by forming or joining a prescribed burning association or cooperative (see chapter 16).

Personal Liability

A common question people ask, whether they are working for a state agency or helping a neighbor, is, "Where does my liability begin and end?" If you are a private landholder, agency, charitable organization, nonprofit corporation, or for-profit business, the legal doctrine of *respondeat superior* applies. *Respondeat superior* states that the negligence of the employee is imputed to the corporate entity if the employee was acting within his or her scope of responsibility and authority, and if there was not a willful and wanton act to injure another (van der Smissen 1990). An employer should and must be accountable for its employees' actions. On the other hand, if the employee's actions are found to be negligent, then the negligence is imputed to the agency, as well as to the employee. The doctrine of *respondeat superior* can be voided by an employee's actions if the employee's negligence is not imparted to the entity but the employee "stands alone." At this point the employer is not liable, and the employee must rely on his or her own personal liability insurance to pay for damages.

An employee can choose three primary actions that will result in the employee "standing alone" or being personally liable for damages caused by a prescribed fire. In an *ultra vires* act, the employee acts outside the scope of authority and responsibility. The negligence of the employee is not imputed to the employer when the employee acts beyond his or her scope of employment. Willful and wanton conduct is the actual or deliberate intention to cause damages; if not intentional, the conduct shows an utter indifference to or conscious disregard for the safety of others. Gross negligence is failure to exercise the care that most people would use, or a person coming just short of a reckless disregard for the consequences of his or her actions.

The first consideration an employee should have before conducting a prescribed burn is to determine if it is within the scope of his or her duties or in the job description. If not, then the employee will "stand alone" and assume all liability. If your supervisors support conducting prescribed burns, ask them to rewrite your job description to show that you can conduct prescribed burns, and the doctrine of *respondeat superior* will protect you.

In some states, agency personnel are not allowed to actively participate in prescribed burns; they can only write burn plans and give technical assistance. Some employees will take annual leave on the day of a burn to assist the landholder. At this point the employee is acting as a private citizen and will "stand alone" in assuming any liability. If in this situation, you should be careful because landholders may still look to you for expertise and may think you are acting in accordance with your job. Either make sure that the landholder knows you are there on your own time and not as employee of any entity, or do not attend the burn at all.

Some disturbing information has been coming out of the federal agencies that use prescribed fire as a land management tool. Some federal agency supervisors have told their employees to get personal liability insurance polices, causing some employees to reconsider their prescribed burning activities. This is a substantial setback to the federal prescribed fire program, because the government needs to compensate for decades of fire suppression and should be using prescribed burning whenever possible. Requiring employees to have personal liability policies may be a risk-elimination tactic that some of the anti-prescribed-fire supervisors and administrators are using. This also goes against what the doctrine of *respondeat superior* states, that if an employee acts within the scope of employment and does not commit any of the three primary actions that cause an employee to "stand alone," then the employee will be protected from personal liability.

Even though employees are protected by the doctrine of *respondeat superior,* they can still be named in a lawsuit. For example, the legal counsel at Oklahoma State University has advised employees that they are protected as long as they are working within the scope of their employment, but they could be named in the lawsuit, at which time employees should retain a lawyer to remove their name from the lawsuit. The university will not help with the cost of the lawyer, and the employee should not trust the employer to provide any assistance. In some cases the lawyers for the employer may attempt to prove *ultra vires* to try to shift the negligence and protect the corporate entity from further litigation.

Another question that people ask is, "What is their personal liability when people are volunteers or assist their neighbors on a burn?" It has been shown that if a person is acting on behalf of an entity, whether paid or not, the person is considered an agent of the entity, and the volunteer's negligence is imputed to the entity under the doctrine of *respondeat superior* (van der Smissen 1990). But liability will still rest upon the volunteer if he or she acts in a negligent manner, just as it will for a compensated employee. To protect both the landholders and volunteers, the volunteers should have the same types of credentials, orientation, training, and supervision as paid employees (Stanton 1995).

Prescribed burning laws and polices are frequently changing, so be sure to keep abreast of all current legislation and changes in agency guidelines. If prescribed burning laws in your state are not adequately written, work with your state and local representatives to change them. Be a positive voice, and provide solid factual information to the people responsible for writing and overseeing fire laws and guidelines within each state. As a result, conducting prescribed burns can be more easily facilitated and risk to the burner and surrounding community will be kept to a minimum.

CHAPTER 3 Prescribed Fire Policy

Ye shall kindle no fire throughout your habitations on the Sabbath day.

—Exodus 35:3

A prescribed fire policy is set by a specific entity and lists the requirements and guidelines that should be followed when conducting a prescribed burn. Policies can include the educational background, training, and experience requirements for personnel; weather and climate conditions under which burns will be conducted; and equipment required for conducting the burn, as well as a statement for dealing with the public. Policies are as varied as the people who write them. Some policies are very simple and friendly toward the people conducting the burns, while others are complex and make performing a prescribed fire next to impossible.

Crisis and emotion have the strongest influence on a prescribed fire policy. A crisis will always bring about some scrutiny of an existing policy or cause those without a policy to write one. Crisis can be defined by a number of factors, such as an escaped fire, smoke management problems, or some type of litigation being brought against the burner or burner's employer. Emotion can influence policy, and many times the media influence public emotion. The emotion may stem from a story relating a problem that occurred during a prescribed fire or may come from a human-interest story about the positive effects of fire. How emotion impacts policy will depend on the type of story and outcome of the situation. Smokey Bear is a perfect example of how an advertising campaign can impact the public's perception of fire. This one story has affected public perception of fire nationwide due to a bear cub surviving a wildfire, which stirred human emotion. The majority of the public considers all fires to be bad because they do not know the difference between a wildfire and a prescribed fire; thus, a cartoon bear has had an impact on prescribed fire policy for years. The emotion from an injury or death of a crew member or outside person can also cause drastic changes in policy, even to the point of discontinuance of the burn program.

An example of crisis and emotion causing a policy to be written occurred at Oklahoma State University in 1996. At that time the Rangeland Ecology and Management group there had been conducting prescribed burns to carry out research and to achieve specific land management objectives for over twenty years. They did not have a written policy in place but conducted all burns with a written fire plan. The spring of 1996 was very dry, with the state imposing burn bans several times during this period. Dry weather and several severe wildfires that had burned large areas throughout the state had put people on edge about any type of fire. The Rangeland group had conducted a research burn on a day when the weather was good for burning, but on the following day the weather became very warm and dry. An area that had not burned on the previous day rekindled adjacent to the firebreak. Personnel were patrolling and mopping up, but they could not prevent the fire from escaping and burning about 120 acres (49 ha) in the process. This was not the major problem: the crisis created was the sense of panic that the fire created in a housing addition 2 miles (3.2 km) away. Local police went through the neighborhood with a loudspeaker and urged everyone to evacuate their homes even though residents were not in any danger. Furthermore, law enforcement officials had not been told to evacuate residents. Both the smoke and the actions of law enforcement terrified a large number of homeowners. Many were so concerned that they loaded their valuables into their vehicles and left, although some stayed to protect their homes. A picture on the front page of the local paper showed one resident on his roof setting up sprinklers. What would have happened if he had fallen off the roof while doing this?

The Rangeland Ecology and Management group salvaged their burn program through a lot of work and public relations, but not before they were forced to have a written policy in hand prior to any further burning. The dean of the division and the president of the university approved the policy. For several years following this crisis the dean requested notification before every burn. This crisis caused a fire policy to be written, because without the policy the burning program could not have continued.

The combination of crisis and emotion can deal a death blow to a prescribed fire program. Thus, having a written fire policy and a public relations plan in place before you begin burning is extremely important for the continuance of most prescribed burning programs. Then if problems do arise, you can show that you already have a written policy addressing the issue at hand. Fire policies are also a good public relations tool and can redirect many of the concerns the public has by showing the benefits of prescribed fire. This takes some of the emphasis off any crisis that has occurred because you can demonstrate that you have been proactive in your actions.

For the above-mentioned reasons those involved with prescribed fires, such as agencies, corporations, and private landowners, should have some type of guidelines or policy in place before burning. Prescribed burning associations are a good example of private landowners cooperating to have a written policy to increase the safety of prescribed fires. Most all of the prescribed burning associations in Oklahoma and Texas have a written policy stating their guidelines for conducting safe prescribed burns. These guidelines are important for the continued operation of the association, as well as for the protection and safety of each individual member involved (see chapter 16).

For other reasons you or your group should have a written prescribed fire policy in place before you begin burning. The most important is to assure the safety of the burn personnel and surrounding public. The policy might address a number of things, such as the safety requirements for personnel, including clothing and training; the smoke management requirements for burns in order to reduce any off-site impacts or problems; steps to take in case a problem does arise; and emergency procedures and contact information. Having a written policy that outlines how situations will be handled can be a great benefit to the safety and peace of mind of the surrounding neighbors and communities.

The second reason for having a policy in place has to do with liability concerns. A written policy will let individuals employed by another person or agency know where they stand in case of an accident or an escaped fire. Before employees burn for an individual or agency, they should know what the employer's policy is concerning liability. If employees act outside the scope of their employment by conducting a prescribed burn, then the individuals are liable for their actions. If burning is part of an employee's job and part of a written policy, then the individual is covered by the legal doctrine of *respondeat superior,* and the employer is responsible for the actions of the employee.

An insurance company may require a person or company that it is insuring to have a written policy in place before the insurance coverage will become active. It is important for you to have a policy in place in the event that litigation is brought against you or your agency. The policy should help prove in a court of law that you understand how to properly conduct prescribed burns. (For further information on liability concerns, see "Managing Liability" in chapter 2.)

It is also important to have a policy that addresses what information is needed for both employee and supervisors. Your policy should include a chain of command, or who answers to whom. The chain of command should define who has the final authority for approving the burn plans and who can say "go" or "no-go" on burn days. The policy may also indicate who should be notified before a burn is conducted or if problems arise and should be clear about who is responsible for the burn and the safety of the personnel involved. An agency or corporate entity must also determine who is in charge of public relations. The policy should determine who handles media inquiries before, during, and after burns, as well as how public relations will be handled if there is a crisis.

Another key reason to have a written policy is to define the parameters for conducting a burn. These parameters may include specification of goals and objectives for the burn, along with determining if burning is the appropriate management practice. The policy may specify the weather conditions required for each fire. It could also take into account the amount and type of equipment required, as well as the size of the crew needed for each burn. It might even include pre- and postburn management strategies, along with postburn monitoring of fire effects.

The most important thing to remember when writing a prescribed fire policy is to keep it simple and not put limits on the fire boss or the personnel conducting the burns. Do not set limits or constraints so narrow that they become unattainable. It is best to set only minimum requirements when writing your policy. Instead of including all equipment that would be ideal for the burn, include only the equipment and personnel absolutely necessary to achieve your goals. This would mean setting a minimum number of people and a minimum amount of equipment required on each burn. If you include all your employees as the ideal number of people needed for your burn, but someone is sick or unavailable, you would not be able to conduct the burn because you would be violating your policy. The same is true for your equipment; do not list every piece of equipment you have available as being needed to conduct burns. Again, what happens when a piece of equipment is broken or unavailable for use? You will not be able to burn because your available equipment does not meet the required equipment listed in your policy. Be sure to take all of this into account when finalizing your policy.

The same reasoning applies if you want to set weather parameters or any other conditions within a written policy; be sure to make the range as wide as possible so you do not overly limit your ability to conduct burns. Remember that all burn objectives and units are unique; many areas can be safely burned under more extreme conditions or with less personnel and equipment than can others.

There are many examples of written prescribed fire policies available for you to look at when planning your own policy. Some of these examples are very specific and lengthy; others are short and to the point. Both types of policy can be good in their own way. Many times policies that are not very extensive may be lacking important information, but longer policies may have too much information. A good example of a policy that is short and to the point is the one from the U.S. Army Corps of Engineers, Tulsa District Office. The surprising thing is that this policy is from a governmental agency, not known for having short documents. Even though this may not be the best policy, it is easy to read and understand. The Tulsa District policy follows:

> Recognizing the fact that the use of prescribed burns is a scientifically accepted and viable tool for habitat manipulation and improvement, the following stipulations are required of all prescribed burn activities on Tulsa District lakes.
>
> All prescribed burns must have a designated fire leader.
>
> The fire leader must have satisfactorily completed a district approved prescribed burn training course before conducting prescribed burns.
>
> The fire leader must submit a burn plan to Tulsa District Office, Operations Division for approval before conducting a prescribed burn.
>
> Burn plans will be kept on file at the lake office for minimum period of three years. (Hogue 2000)

Some very lengthy written policies are currently in use. The National Park Service's Wildland and Prescribed Fire Management Policy has nineteen chapters and covers numerous subjects pertaining to prescribed and wildland fires.

Some of the topics include safety, training, plans, fuels management, and monitoring (National Park Service 1999). The Nature Conservancy (TNC) also has a lengthy fire policy. The TNC manual extensively covers prescribed fire and is divided into four sections: (1) Introduction; (2) Administration, Personnel, and Insurance; (3) Requirements and Guidelines; and (4) Fire Management Process (Seamon 2004). The more personnel and land area an entity manages, the more specific the policy must be.

Therefore, you want to be specific when writing fire plans and policies, but, know that problems could arise later if your guidelines are too narrow. Policies should also be written as broadly as possible to avoid any potential problems. The main thing to remember is that each fire is different and requires a different set of weather conditions, number of personnel, and amount of equipment. Your policy should be written broadly enough to cover all the vegetation or fuel types that exist where your organization conducts burns.

CHAPTER 4 Public Relations

Every tree that bringeth not forth good fruit is hewn down, and cast into the fire.

—Matthew 7:19

A good public relations plan is just as important as a good prescribed fire plan. Working with the public, instead of ignoring them, will benefit a prescribed burning program many times over. Remember that concerns about public relations are not as important in more sparsely populated areas as they are if you want to conduct a burn next to a major city with civic groups, media, and others watching over your shoulder. Also, public relations issues can be minor for private land managers compared to those for personnel trying to burn on public lands. So, evaluate each situation and plan accordingly.

Public relations is defined as a management function that evaluates public attitudes, identifies the policies and procedures of an individual or an organization with the public interest, and plans and executes a program of action to earn public understanding and acceptance (Seitel 2001). It can also be described as the methods or activities employed to promote a favorable relationship with the public. In some regions public relations for prescribed fire is, or should be, a full-time job. A study conducted in the South found that public opinion was the greatest barrier to implementing prescribed fire on private and state forest lands (Haines and Busby 2001). It can be very difficult to overcome the negative image of fires that the media and Smokey Bear present to the public. You should follow these four steps for a successful public relations campaign (Jacobson 1999): define the objectives, identify your audience, develop and implement your strategy, and evaluate the results.

Be clear about your objectives (i.e., public acceptance to prescribed fire) and how you are going to get your message to the masses (media, talks, etc.), and determine your target audience. If you have not determined these items, then you are wasting your time and will not be able to educate the public. You also need to have a well-thought-out strategy and contingency plan in place before beginning your fire program. Addressing these factors before you begin your prescribed burn, and then using them as a guideline after the burn, ensures that you will be able to easily evaluate your results.

Your Public Image

Public image or public perception is a very important part of a public relations program (Figure 4-1). Public perception or image may not be what we really are but how we are perceived through the eyes of the public (Inter-Agency Pre-

Figure 4-1.
A positive public image is important for maintaining an active prescribed burning program. Attitude, professionalism, and proper equipment are the three aspects of a positive public image (photo courtesy of Stephen Winter).

scribed Fire Course 1999), and it is very important to have a positive public image. Your prescribed burning program may be the best one in your particular area or agency, but if the public in your area does not perceive that, your burning program will be in jeopardy. The window through which the public views you should remain transparent so the public does not feel that you are hiding something from them. This is very important for the continuation of your prescribed burning program. Unless you have a positive public image, the support from your supervisors or the public will falter, which may cause unfavorable media attention and, in turn, cause you to possibly lose financial and other support for the entire burning program.

There are three aspects of a positive public image: attitude, professionalism, and proper equipment (Inter-Agency Prescribed Fire Course 1999). Attitude is the tendency to think in a particular way about a specific subject; people will have a positive, negative, or neutral attitude toward particular issues depending on their prior exposure to them (Jacobson 1999). Unfortunately, most of the public's attitude about fire has been shaped through misinformation or lack of information. Attitude relates not only to what the public believes about prescribed fire but also to the attitude that the person(s) conducting the prescribed fire has toward the public and others inside the organization. Many times there is a sense of mystery about working with prescribed fire, and the public will often look on in awe at what has been accomplished because it is a process about which they know little and perceive as controlling an element of danger. Sometimes prescribed burners can begin to believe that the work they are doing is of a greater purpose than anything else. This arrogant attitude does not influence many people outside of their peers but can create animosity and negative feelings between burning programs and the public.

The second aspect of a positive public image is professionalism. Professionalism relates to your ability as burn boss, fire management officer, or land manager to properly conduct prescribed burns and the knowledge that you bring to each burn area. This is for the benefit not only of your peers but also of the people your prescribed fires are impacting. Professionalism also includes your physical appearance. Do you appear timid and uncertain when approached with questions about your burning program, or do you look people in the eye and explain your burn program in a professional manner? The public will watch you and your crew from a distance to see how you behave, react, and handle yourselves in everyday situations when you are not conducting prescribed burns. If you show people that you are professional in all aspects of your life, then the public is more likely to believe that you are capable of conducting and managing a prescribed fire.

The third aspect is proper equipment, which also relates to professionalism. Proper equipment is a tangible projection of how prepared you are for the burn and of the amount of protection you can provide for the surrounding landowners and community. Imagine how the public might react if they see you and your burn crew pulling up to a burn unit with vehicles and equipment that do not work. The public might well feel that your program is not trustworthy, a public image that can doom your prescribed fire program.

Public Perception of Prescribed Fires

Before you start interacting with the public, it is necessary to assess your public image, even if you have been burning for several years. You can accomplish this in several ways. First, mentally step outside your fire program and try to look at yourself through the eyes of the public. Hopefully you will be able to get an objective view of your program and see which areas are working well and which areas may need improving. Next, you need to know what the media is saying or writing about you and your fire program. You should pay attention to what friends and adversaries are saying about your prescribed burning procedures. Public input is important in identifying what levels of prescribed fire are socially acceptable (Weldon 1996). Finally, you should be honest with yourself and others about any changes that need to be made. This can be the most difficult part because it is often hard to see problems within your own organization.

Prescribed fire managers run into numerous barriers when trying to deal with the public during the introduction or implementation of prescribed fires (Jacobson 1999). Most of these problems are just a misunderstanding and lack of education on the part of the public but illustrate the need for a positive and effective public relations program for everyone who conducts prescribed fires.

The following are some of the problems the public has with prescribed fire (Taylor and Mutch 1985; Daniel 1990):

- Misunderstanding on the part of the public and agency personnel about the ecological effects of fire, coupled with inadequate feedback showing the positive results of prescribed fires
- Public fear of fire and lack of public information about smoke, danger, and the immediate effects of prescribed burning
- Adverse effects of fire on public aesthetic and recreational values of the land, and a lack of public preparation for postfire vegetation and wildlife recovery
- Lack of public confidence or trust in agencies because of inconsistent policy or previous fire management tactics
- Successful Smokey Bear fire prevention campaign that portrays fire as destructive and dangerous
- Media misinterpretation of prescribed fire practices
- Disjointed information due to lack of interagency coordination and cooperation in fire awareness programs
- Lack of a prescribed fire information program that is effective, coordinated, and targeted to key audiences

While working on reintroducing fire into the Bitterroot Valley of Montana, fire managers reported that strategies, goals, and objectives for involving the public should be a part of a prescribed fire restoration program, and that the challenges of implementing prescribed fire could not be resolved without interacting with the public and communities that are affected. They also stated that land managers must talk, listen, and respond to the public. Four categories of information about prescribed fire should be brought to the public's attention: ecological facts about fires, as well as the benefits, consequences, and risks involved with implementing prescribed fires (Weldon 1996). If a land manager can develop a simple and effective way to present this information to groups and individuals, many problems can be easily avoided.

Presenting the ecological facts about a region can be accomplished by describing how the country looked prior to settlement or by using old photographs and comparing specific areas over time. Finding the photographs can sometimes be difficult, but by contacting local historical societies or civic groups, you can work with other individuals in the community. And outside people becoming involved and having a vested interest in your project will help strengthen your program.

Field trips or field days to areas that have burned are a good method for presenting ecological facts to the public and for allowing the public a chance to see actual postfire effects. Demonstration plots or areas can also be very useful because they allow people to see and touch areas that have had different fire regimes used on them. Oklahoma State University has started using demon-

Figure 4-2.
Demonstration areas are useful tools for showing people how fire affects plant and animal communities. Here land managers are viewing the effects of fire frequency on sand shinnery oak grasslands. Small demonstration areas like this can impact the management of thousands, or even millions, of acres of land.

stration plots to show the effects of fire on plant communities burned at different times of the year. These season-of-burn plots have been placed in different vegetation regions throughout the state and have had a profound impact on the people who visit them, mainly because visitors are able to walk through the plots and see for themselves the effects of a fire conducted in December compared to one conducted in July (Figure 4-2).

Other methods of disseminating information include the use of publications and media features, such as pamphlets, fact sheets, and news releases. Remember to present the information in terms that your audience can understand.

The benefits of prescribed fire are probably the easiest topic for most fire management personnel to discuss and demonstrate. Numerous books, journal articles, and research papers describe the benefits of fire to plant life, wildlife, livestock production, timber production, wildfire reduction, and water quality and quantity, as well as benefits to homeowners in the wildland-urban interface. When describing these benefits, make sure to use examples that relate to the group you are visiting with, and keep the information on a level everyone can understand. Remember that it is not always the most obvious benefit that will stir people's emotions; more often than not the little things, like the benefits to butterflies or songbirds, are what changes an individual's or group's attitude toward prescribed burning.

Describing the consequences of *not* using prescribed fire is a key message to help the public understand the benefits of prescribed burning. Relating the detrimental impact of fuel buildup due to lack of fire, as well as the possible emotional, physical, and economic impacts of a wildfire to the community, can also be useful. Describe the trade-off between a wildfire and a prescribed fire; this could include the comparative amount of smoke produced, its impact on human health, and safety issues. Discuss how a wildfire would impact an individual's life and property. Explain the risk and cost involved with suppressing a wildfire compared to the risk and cost from a properly conducted prescribed fire. Always present the public with the facts so they can make a well-informed decision.

Finally, you should always be honest with the public and explain the risks involved with implementing a prescribed fire. Describe your goals and objectives for conducting the burn. It is also important to explain under what weather conditions the burn will be conducted and why certain conditions are better than others for burning. Explain the difference between weather that is favorable for wildfires and weather prescriptions for a prescribed fire. Talk to them about the risks for spotfires and about what will happen if the fire escapes. Tell them how you plan to reduce this risk, and let them know about the experience that your crew has with previous prescribed fires, the type of equipment available for you, and how you will suppress the escaped fire if one does occur. Remember that the only fire most people know about is wildfire, and they believe that all fires behave in the same way.

You can use several methods to relay your message to the public, and although there is no single, perfect method, you can use a combination to create a program that effectively meets the needs of the public. It is important to change your method of informing the public based upon the focus of the group you are currently addressing. When addressing the public, you need to remember that communication is most persuasive when it comes from several, highly credible sources (credibility = trustworthiness, expertise, and power); is a simple message; is easy to understand; relates to the audience; and arouses personal relevance and involvement (Jacobson 1999).

Talks, Forums, and Public Meetings

Using talks, forums, or public meetings is an effective way to reach larger audiences and target specific groups. When preparing talks, remember to use facts that are easy to understand and that relate to your audience. Forums can be either positive or negative opportunities, depending on how you approach them. Find out who is in charge, what the format is, and who the other speakers are. Be careful because sometimes you may be set up as the scapegoat to an antagonistic crowd.

Most land management personnel are terrified of public meetings, mainly because of negative past experiences or horror stories from colleagues. Public meetings can be a very positive experience for both parties if they are properly conducted and have a qualified moderator. It is best to bring in an outside person or someone whom the public views as neutral to moderate the meeting.

Having an outside moderator allows land managers to concentrate on presenting their information instead of dealing with unruly participants, which could possibly cause unwarranted animosity from members of the crowd.

I was involved with a prescribed burn conducted on U.S. Army Corps of Engineers (USACE) land and managed by the Oklahoma Department of Wildlife Conservation (ODWC). Both managers of the respective agencies had limited prescribed fire qualifications, and the USACE personnel had had very bad previous experiences with public meetings. The two agencies asked us to participate as the experts at a public meeting to discuss the fire plan and the benefits of prescribed fire. Also in attendance was the assistant fire marshal for the city of Edmond, Oklahoma, because the burn was within the city limits, and homeowners lived next to the burn unit. The meeting went very well and was beneficial for both sides for two reasons: (1) it was very well planned, and (2) it was moderated in a professional manner by an outside source.

All public meetings should have a set of ground rules to facilitate the discussion. There should be a moderator, a person who is qualified and can handle himself or herself and the crowd in a professional manner while maintaining control of the meeting. You should have an agenda and stick to it; do not allow people to change the subject or get off track. Let the experts speak first; often this will answer many of the questions the audience has. After the experts speak, allow the people who are affected to voice their concerns and allow the experts time to answer any questions. You should have a sign-up sheet for people who want to speak. If you make the speakers sign up, you will know how many people want to talk and the moderator can determine beforehand how long each person can speak. Also, if you make people sign up before talking, it may limit who talks, as some people will not sign up because they want to remain anonymous. If possible, after the meeting let the public visit with the experts one-on-one. Visiting with an individual is one of the most effective ways to influence a person's feelings about a subject (Figure 4-3).

The speakers should have a set of ground rules as a guide. The following is a list of the ground rules that we used for the speakers at a meeting in Edmond, Oklahoma. This list was handed out to each person at the meeting and was compiled by "Community Connections" for the City of Edmond, who provided the moderator and set up the meeting.

- One person speaks at a time.
- Tolerate no interruptions.
- Speak from experience, not gossip.
- Do not allow put-downs of self or others.
- Do not allow side conversations (distractions).
- Treat everyone with respect.
- Be respectful of differences of opinion.
- Make space for others—do not dominate.

Figure 4-3.
Public meetings can be an effective way to present prescribed fire programs to community members, but you must have a qualified moderator, adhere to an agenda, and determine ground rules for the speakers to follow.

- Do not make attacks (verbal or physical).
- Stick to the agenda or project.
- Smile.

These rules should be handed out and read to the audience by the moderator before the meeting begins so that everyone in attendance understands them. You should not allow any deviation from this list of rules, and people who do deviate should be asked to leave the meeting. If you expect unruly people at the meeting, it is best to have uniformed law enforcement personnel at the meeting. Many times their presence will stop most outbursts before they begin.

Tours, Field Trips, and Demonstrations

Tours, field trips, and demonstrations are other methods for making personal contact with your audience. These are time-proven methods, but remember to keep the tours interesting and flowing so that people will not become bored. Anytime a hands-on approach can be used, it will be more effective. Set up demonstrations that people can walk through or touch, keeping in mind that tours, field trips, and demonstration sites should have attracting power (Will people stop and look?), holding power (Does it keep a person's interest?), teaching power (Will people learn from it?), and motivating power (Will the people be motivated to find out more or take action?) (Bitgood and Patterson 1987). If you keep this list in mind when you set up your tour or demonstration, you will increase your likelihood for being successful.

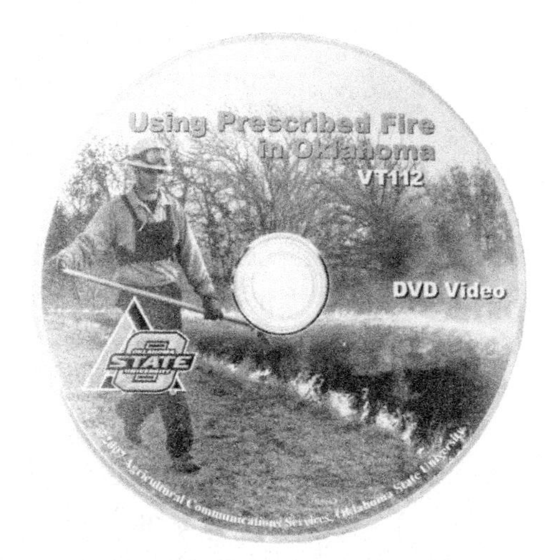

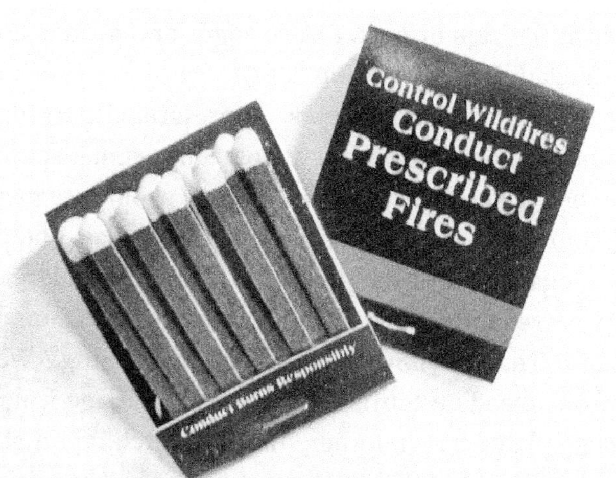

Figure 4-4. Numerous methods are available to present your message to the public: for example (clockwise from top left), publications that explain the benefits of prescribed fire, videos promoting the use of fire, bumper stickers with easy-to-read messages, and novelty handouts such as these matches. Be sure to use a short message that is easy for your audience to understand.

Literature, Pamphlets, and Fact Sheets

Written material can be a good way to disseminate information about your burning program. Numerous publications are already available from federal and state agencies, universities, and nonprofit organizations that use or promote prescribed fire (Figure 4-4). The National Park Service (1983) has found many advantages to using written publications: portability, ease of design, relatively low production costs, and souvenir value. Also, they can be read at an individual's own pace, provide a source of reference, and can be revised easily. At the same time the National Park Service also lists some limitations in using written publications: the text may be lengthy, the reader has only the text to create an image of your organization, written information requires effort by the person reading and may need periodic revision to remain accurate, and public interest may wane if people have to read a publication. Keep this information in mind when preparing or using written publications for public relations campaigns.

Letters, Signs, and Posters

Letters are good for reaching specific groups of people. When writing letters to people impacted by a prescribed fire, be sure to answer the questions of who, what, where, when, and why. Make the letter brief, informative, and easy to read; try to relate the message to the specific audience. Letters can have drawbacks, such as having to obtain your audience's addresses, cost of mailing, and making sure people read the letter.

Signs and posters can be used to provide a simple, short message. This type of public communication does not allow for much detail in the message, and people frequently only glance at the sign or poster. Also, signs and posters do not allow you to target a specific audience (Jacobson 1999).

Another method that can be used to spark interest in prescribed burning programs is to produce bumper stickers. Again, they allow only a small space for a short message. Be sure the sticker is easy to read because people are just going to have a few seconds to read it. Using bumper stickers may also help you develop a catch phrase or slogan for your campaign. In Oklahoma we developed a bumper sticker that is passed out free of charge and states, "Oklahoma Is Native America; Let's Keep It That Way! Support Prescribed Burning." We used the "Native America" from the state tourism slogan and wordage from Oklahoma's automobile license plates. The bumper sticker is catching on all over the state; hopefully it will help spread the word about prescribed burning (Figure 4-4). A similar bumper sticker in Texas looks like the state flag with the words "Prevent Wildfires; Support Prescribed Fires." This is a good slogan that makes a positive statement for prescribed burning.

Press Releases

A press release or news release is a written article describing your program that can be used by newspaper, radio, and television to reach the public. Be sure the headline and lead sentence will grab the attention of the editor or reporter who reads them. Here are ten tips that will help you write effective news releases (Jacobson 1999):

- Know what you want to say.
- Ensure that the story is newsworthy.
- Keep it simple and direct.
- Keep it brief.
- Emphasize the human interest.
- Be timely.
- Establish the credentials of your organization.

- Follow the standard format.
- Provide a picture.
- Write well.

If you have limited experience writing or believe that your writing skills are not adequate, seek some advice or help from people who have experience with this type of writing. Remember that the public will use this press release to form an opinion of your organization or program. Therefore, you want to present a good image through your writing.

Interviews

Interviews with the news media can be an effective way to reach the public. When working with a reporter, be prepared and anticipate the questions you might be asked. Most reporters ask the five *W*'s: who, what, where, when, and why (Jacobson 1999). You will want to have all of your facts and figures ready and be prepared to answer questions, give examples, and cite specific information about your prescribed burning issue and organization. You should also be prepared to answer questions about subjects that are controversial or that may expose problems in your program, although you will not want to dwell on these negatives. Be sure to reiterate the positive aspects of your program and share examples or experiences so the audience can relate to the information.

One-on-One

Most people understand information best when it is presented to them in a one-on-one situation. Although the information reaches only one person, it could possibly impact many others when this one person talks with friends or coworkers. Remember the most important aspect of a one-on-one discussion is to listen more than you talk. A lot of the time people just want to tell someone their concerns. Once they voice their anxieties about fire, you can begin to explain how your fire program will address each of these fears. If this person is convinced, then he or she will tell others, hopefully causing a chain reaction that will spread the positive word about your burning program. This can also backfire if your audience is not convinced, so be professional and courteous to everyone you visit with to avoid giving people a reason not to trust your program.

Working with the Media

If you plan to use a public relations campaign, you are going to have to work with the media to help get your message out (Figure 4-5). There are several things you should never do when working with the media, for example:

- Do not give or expect favoritism from the media. This will possibly cause the other media personnel to become disgruntled with you and your

Figure 4-5.
Working with the media is an important part of public relations. Be prepared for reporters' questions, and answer in short sound-bite-type responses.

program, and they may report unfavorable items or quit covering your program altogether.

- Do not ever attempt to talk "off the record"; nothing is ever off the record.
- Do not make offhand remarks. These will often come back to hurt you later.
- Do not hide behind technical jargon; it may sound professional, but remember who your audience is and think about whether or not they will understand you.
- Do not disappear when unfavorable news breaks. Problems may be difficult to face, but is better to be the one presenting the facts than to have a reporter go to an "expert" who knows little about fires or about the facts of your particular burn.
- Do not stonewall reporters; this will also cause them to seek their information somewhere else or cease to cover your fire program altogether.
- Do not try to fool the press. Remember that they were hired in their respective field because of their knowledge and skills; most are educated and know how to research the facts about a story, so be straightforward and truthful with them.
- Do not tolerate any openly belligerent behavior from a reporter. If this happens, talk to the reporter's editor and ask to have another reporter cover the story.

- Do not try to be a news personality; this will only make you and your burning program look substandard.
- Assume that media personnel are professionals in their particular field, just as we are, and treat them with the same respect that you would like for people to give you.
- Empathize with reporters' needs; they are only trying to do their jobs.
- Know with whom you are talking when you give interviews. Ask reporters whom they work for and why they are here. This information can help you determine the best information to present and the best way to present it.
- Give equal consideration to all types of media, both written and visual.
- Consider if the event (prescribed fire) is worthy of local media coverage only, or if it should have regional or national coverage.
- Provide the complete story to the media; they are paid to find out the news, and it looks better if the information comes directly from you.
- Take the initiative in relating bad news. Do not make the reporters come to you for the information because it may appear that you are trying to hide it from them.
- Pay attention to dates, times, places, and spelling, and make sure that reporters are receiving the correct information.
- Be prepared for all possible questions, and make sure you know the facts. If you think that the media will ask you about specific numbers, such as total acres burned or number of people involved, write it down. This will keep you from stammering in front of everyone and looking like you do not know what you are talking about or like you are hiding something.
- Be punctual for meetings, and be sure to return calls as soon as possible.
- Keep your appointments and your promises, and be positive and courteous when visiting with media personnel.
- Document your contacts with the media. This will help you keep a record of whom you talked with and when you spoke with them.
- Monitor the news programs and newspapers you have talked with to make sure their reporting is accurate according to the information you have given them.
- When presenting information, remember that conflict and tragedy are newsworthy.
- Have one or two key messages that you want to present, and learn to communicate the message in ten- to fifteen-second sound bites.

It may be important to have a specific media relations person during a prescribed fire. The fire boss is normally too busy to answer questions during a fire and should not be distracted from his or her appointed task of managing the burn. You should also inform everyone on the burn crew to direct all media questions to the contact person. This can save a lot of problems in the long run. Reporters may ask the same question to five different crew members and receive five different answers, or a disgruntled crew member may spread the wrong information in order to discredit you or the program. This does not help a public relations campaign. Be sure to have only one person designated to be the spokesperson, and have all media inquiries go to that person. Make sure the spokesperson is competent, informed, and courteous. Remember that even though agency policy may dictate how and when to disseminate information, you are still the messenger and may have to face some hard questions if a problem occurs.

Dealing with the "Real" Public

While working with the public on wilderness fire policies, a group of fire managers made an important statement about dealing with the "real" public. They state, "Although the public now is more conceptually knowledgeable and supportive of prescribed fire practices, the fact remains that many people still react negatively toward the direct results produced during a fire event." These differences between public *conceptual* and *perceptual* response are at the heart of the fire manager's dilemma: "You can hold public meetings, explain your fire policy and program, and get strong support from the community. But when the smoke from your prescribed fire hits the town, you had better be ready for some very different reactions" (Taylor and Mutch 1985, p. 49).

This is very true; I have been involved with public reaction where there was support when we began discussing the burn with the community, but the people's reactions changed when the smoke became visible. While working on the burn with the USACE and ODWC in Edmond, our public meeting went very well, and during the meeting we had no negative comments or objections. On one of the days we were planning to burn, the phone to the Edmond Fire Department and USACE office rang all day long with concerned people wondering what was happening. Where were these people during the planning and public meeting stage? You will need to be prepared for this type of reaction and have a plan in place to deal with it.

It is important first to determine the goals and objectives of a public relations campaign and then determine the best method to get your message to the public. Remember that a good public relations program is an all-day, every-day job, and it does not stop when five o'clock rolls around.

CHAPTER 5 Fire Weather

Weather is the most important factor in deciding whether or not to conduct a prescribed burn. This is due in part to the weather's dynamic nature and the difficulty in making accurate predictions. Weather parameters such as temperature, relative humidity, wind speed, and frontal boundaries have drastic influences on prescribed burning and fire behavior. Fire behavior characteristics such as intensity and rate of spread are directly related to weather. The weather also has bearing on how much smoke is produced, how well the smoke lifts, and which direction the smoke travels. Therefore, being able to understand the weather is a very important part of conducting prescribed burns.

A flame devoureth before them; and behind them a flame burneth: the land is as the Garden of Eden before them, and behind them a desolate wilderness; yea, and nothing shall escape them.

—Joel 2:3

Temperature

Temperature primarily affects prescribed fires by heating or cooling the objects to be burned. Warmer fuel particles require less heat for combustion, while cooler particles require more heat. There are times when higher temperatures may be required when conducting a burn to meet objectives; if a burn unit has large quantities of fine fuel, it may be safer to burn with lower temperatures to reduce fire intensity.

The following local factors can affect temperature:

- *Season of year:* Longer amounts of daylight increase the heating of the air and cause higher surface temperatures.
- *Cloud cover:* Clouds can reflect and absorb solar radiation, causing a reduction in the amount of heat available at ground level. Clouds can also redirect solar radiation to the earth's surface, causing a warming or insulating effect.
- *Topography:* Topography causes differences in temperature at specific locations due to the variation in the angle at which the sun's rays strike the land's surface. The angle of the slope impacts surface heating and cooling.
- *Aspect:* East-facing slopes heat up early in the day, while west-facing slopes heat up later in the day. The highest temperatures are normally found on southwest-facing slopes.
- *Elevation:* Elevation and temperature are inversely related: that is, as elevation increases, temperature decreases; and as elevation decreases,

temperature increases. The temperature change ranges from 2°F to 5°F per 1,000 feet (1°C to 2°C per 305 m) of elevation.

- *Proximity to large bodies of water:* Increased moisture from large bodies of water will cause smaller variations in high and low temperatures. Also, the temperature of the water can affect air temperatures.
- *Air mass:* The type of air mass, cold or warm, can rapidly impact temperatures.
- *Shading:* Shade causes a reduction in solar radiation, and thus a variation in surface temperatures. Shading can be caused by clouds, smoke, topography, or vegetation.
- *Vegetation:* Solar radiation is intercepted by vegetation, which influences ground temperatures. Green foliage does not warm up as much as dormant vegetation, and the highest daytime temperatures are found in the upper part of the crowns of trees because leaves trap the radiation.
- *Wind:* Wind causes the transfer of heat from the ground surface to mix with the air. During the day strong winds reduce high surface temperatures, and at night these same winds prevent low surface temperatures.
- *Relative humidity:* Drier air allows for greater differences in daily temperatures, while moist air reduces variation in high and low temperatures.
- *Temperature inversions:* As air cools (especially at night), it becomes heavier and sinks; the temperature variations created at different elevations can cause problems.

Relative Humidity

Moisture in the atmosphere impacts fire behavior by increasing or decreasing fuel moisture, which will affect flame length, rate of spread, energy released, amount of smoke produced, and whether or not the unit will even burn. The two main measurements for the amount of moisture in the atmosphere are relative humidity and dew point.

Relative humidity is the ratio, expressed as a percentage, of the amount of moisture in a volume of air to the total amount that volume can hold at a given temperature and atmospheric pressure. Sources of moisture that impact relative humidity are evaporation from bodies of water and moist surfaces, along with water loss from soil and transpiration from plants.

The main rule of thumb that all personnel conducting prescribed burns should remember is that for every 20°F (11.1°C) increase in temperature, relative humidity is reduced by half, and vice versa. For example, if the temperature is 60°F (15.5°C) and the relative humidity is 80% at 8:00 A.M., and you are planning on burning that morning, what would the expected relative humidity be in the afternoon if the forecasted high temperature is 80°F (26.6°C)? It should be around 40%. Is this going to be within your prescription? Again, if the temperature is 40°F (4.4°C), relative humidity is 40% in the morning, and the forecasted high is 60°F (15.5°C), what should the predicted relative humidity be? From the rule of thumb just given, it would be around 20%. Does this fit within your burn conditions? This rule of thumb can help predict relative humidity and fire behavior conditions later in the day.

Relative humidity is normally inversely related to temperature. As temperatures increase, relative humidity decreases. Usually the highest relative humidity is at daybreak when the lowest temperature occurs. The relative humidity will reach a minimum at about the same time the maximum air temperature is achieved.

The following factors affect relative humidity:

- *Topography:* Large, relatively flat areas have similar relative humidity readings, whereas hilly or mountainous areas have large variations within a small area because of uneven heating due to the effects of slope and aspect. Because of temperature decrease, relative humidity will normally increase as elevation increases.
- *Wind:* Wind mixes evaporating water vapor with surrounding air and evens out temperature extremes. This helps reduce place-to-place variation of relative humidity. Location and type of wind will drastically impact relative humidity. In the southern Great Plains a southeasterly wind will bring up moisture from the Gulf of Mexico, increasing relative humidity, whereas a southwesterly wind will bring drier air because the air has traveled over the arid regions of Mexico and the southwestern United States. The same types of factors impact regional relative humidity extremes in other U.S. areas.
- *Clouds:* Clouds affect heating and cooling through shading, which impacts relative humidity. Humidity is normally higher on cloudy days and lower on clear days. Clouds will also bring moisture into the atmosphere because they are made up of water vapor.
- *Vegetation:* The type and amount of vegetation affect the temperature and wind, which are direct factors linked to relative humidity. Relative humidity readings will differ between a closed canopy forest, a savanna, and grassland. The relative humidity will even differ in the forest area due to canopy cover and height variation within the canopy. The humidity will normally be greater within a closed canopy during the day, but will be lower at night than that of nearby openings. Whether the vegetation is growing or dormant will also impact the amount of moisture in the atmosphere.
- *Air masses or fronts:* Air masses that form over bodies of water have higher moisture content than those that form over landmasses. An

example is the air masses that form over the Pacific and move from the northwest into the Great Plains; these air masses normally have moisture within them. Those that form over northern Canada and move down are cold but dry and have lower relative humidity readings. Normally, if a warm, dry air mass replaces a cool, moist air mass, or vice versa, there should be a large change in the amount of moisture in the atmosphere. However, not all moist fronts increase relative humidity, and not all dry fronts decrease relative humidity.

Dew Point

Dew point is the temperature at which the saturation vapor pressure equals the actual vapor pressure (Schroeder and Buck 1970). If the air temperature reaches the dew point, condensation occurs because the amount of water vapor in the air is greater than the maximum amount that can be held at those lower temperatures. When temperatures reach dew point, fuel moistures become higher and combustion of certain fuel classes may not occur. If you are in the mop-up phase of your burn, the temperature reaching the dew point can be welcome assistance by drastically slowing down combustion of material or causing the fire to go out. On the other hand, low dew point temperatures will cause fuels to dry out, making the burn more effective.

During spring months on the southern Great Plains, which is the peak of the burning season, dew point temperatures can be quite high. They may range from 65°F to 75°F (18.3°C to 23.8°C), which are adequate temperatures for severe thunderstorms to develop. Because extreme fire behavior can occur under these conditions, you should exercise care when conducting burns during high dew point temperatures.

Air Masses

There are two types of air masses: low-pressure and high-pressure systems. A low-pressure system, or low, is symbolized by an "L" on a weather map. Low-pressure areas are characterized in the Northern Hemisphere by counterclockwise air rotation. Because the air movement is inward and rising, the air cools the clouds and relative humidity increases. If adequate lift and moisture are present, precipitation will form.

A high-pressure system, or high, is symbolized by an "H" on a weather map. In the Northern Hemisphere high-pressure systems have a clockwise air rotation. The air movement is outward, with descending air flows. Highs typically have minimal clouds and little to no precipitation.

The boundary between two air masses is called a front. This is the leading edge of an air mass and where most of the active weather occurs. Weather conditions that can occur along a frontal boundary are changing, strong, and gusty winds; storms; and precipitation events. The three types of fronts are cold, warm, and stationary. A cold front occurs when a cold air mass replaces a warm air mass; conversely, a warm front occurs when a warm air mass replaces a cold air mass. In a stationary front a frontal boundary does not move or moves very little.

Typically, when a cold front approaches an area, winds from an opposite direction will increase (if a front approaches from the northwest, winds increase from the south), but the winds will then switch to the direction the front is coming from right before or as it passes (if a front approaches from the northwest, winds change to north-northwest). Most precipitation events occur as the cold front passes, but afterward the weather can clear for several days, allowing for good burning conditions. If the air mass has limited moisture, two things can happen when a cold front passes through. Either the front will create exceptional conditions for conducting prescribed fires, with light to moderate steady winds, sunny and clear conditions, excellent lift for smoke dispersion, and pleasant temperatures; or it will dry out the atmosphere too much. When there is very little moisture in the air, fuels become exceptionally dry and can be a source of extreme fire behavior. Therefore, you should exercise caution when burning under these conditions.

A warm front will normally cause a shift and increase in the winds and is normally linked with moist air, higher relative humidity, and precipitation events. A stationary front will have varying weather depending upon which side of the front you are on and the atmospheric conditions present. The main thing to remember is not to burn within twelve hours of any predicted frontal passage.

The determining factor of air mass movement or weather patterns, along with moisture flow, is the jet streams that flow over the United States. These are extremely fast-moving westerly wind flows in the upper atmosphere; in fact, most jet streams have wind speeds ranging from 50 to 200 mph (80 to 322 km/h). The two main jet streams that flow over the continental United States are the subtropical and the subpolar. The subtropical jet stream lifts up from the south and flows across the continent during the summer months. It brings warmer air masses up from lower latitudes. During the winter months the subpolar jet stream moves down and impacts our weather patterns with cooler air masses. If air masses are not pushed along by the jet stream, they will sit in one spot and become what is termed a cutoff low or high. This pattern continues until the flow of the jet stream moves back and pushes the system out.

Hurricanes are another type of air mass that can cause problems during prescribed burns. Hurricanes are low-pressure systems that form over the warm waters of the Atlantic, Caribbean, or Gulf of Mexico. They impact not only the weather along the coast but also the weather for hundreds of miles inland. The potential problems can be a several-day period of heavy rains and flooding or just increased clouds, winds, and relative humidity.

Wind

Wind is primarily caused by the rotation of the Earth, which leads to an unequal heating of the equatorial and polar regions. A secondary cause is the circulation accompanying high- and low-pressure systems produced by the unequal heating of land and water. The greater the variation in temperature between air masses, the stronger the winds will be. The rotation of pressure systems and the direction from which they are coming will have an effect on wind speed and direction. Then, once the leading edge of the system passes, the center will normally have calm or light and variable winds, which are not good conditions for burning. The back side of a system will usually bring milder weather, as well as different wind directions and speeds.

The following factors affect wind:

- *Topography and water:* Topography and water affect winds by increasing or reducing friction, which causes localized changes in speed and direction. Canyons and riverbeds will cause wind speed to increase and will change the wind direction to follow the up- or downslope aspect of the physical feature.
- *Diurnal variation:* Diurnal wind patterns are caused by differences in temperature between night and day, and by the different thermal properties of land and sea surfaces that cause them to heat and cool at different rates. For this reason, winds become lighter and more variable at night in most areas.
- *Vegetation:* Elevated vegetation, such as trees, may lift the wind up and above the area to be burned. The friction loss will depend on the architecture and density of the trees. The winds will be higher above the canopy and somewhat reduced under the canopy. Vegetation that is short in stature, such as shortgrass prairie, will have little friction effect on wind speed.
- *Eddies:* Formation of eddies on the lee, or downwind, side of physical features can be a problem when burning. Eddies are caused by objects in the path of the wind, such as hills, mountains, and trees. The impact of the eddy is determined by the size and shape of the object in the path of the wind, along with wind speed and direction. Burning near these types of features can cause localized increases in wind speed and variable wind directions, so you will need to be extra cautious. Wind eddies can also cause hazardous situations on the fireline, such as changing fire behavior and intensity, as well as cause spotfires.
- *Frontal boundary:* The frontal boundary is the zone of change from one air mass to another. Wind direction can change 45° to 180° from one side of the boundary to the other. Wind speed will depend on how fast

the system is moving and the differences in temperature between the two systems.

- *Foehn winds:* Foehn winds are local winds, usually warm and dry, that are found in mountainous regions. They affect the local weather differently and are known by different names, such as Chinook, Santa Ana, or Mono winds.
- *Slope effect:* Slope effect is the effect on winds in mountainous areas that is caused by heating and cooling patterns during the day. Anabatic winds will flow upslope during the day due to heating, and katabatic winds flow downslope as surfaces cool in the evening. This same pattern is found in valleys.
- *Thunderstorms:* Care should be taken when burning on days favorable for the formation of thunderstorms, as several problems can arise. When thunderstorms form, the convection or uplift causes winds to flow in toward the storm. The result is direction shifts and increased wind speeds. Many times a gust front may move out from these storms with extremely strong and dangerous winds. Another problem can arise if a thunderstorm begins to develop but collapses instead. This can cause a gust front to move out from the disintegrating storm with strong winds that can travel great distances. You may not have been able to see the storm when it was trying to form, but you will definitely be able to feel its effects, so exercise extreme care when burning on days when thunderstorms can develop.
- *Sea breezes:* These winds occur in close proximity to large bodies of water. They are formed during the daylight hours as the landmass warms to temperatures greater than those of the nearby water. The warmer air rises and expands, causing the cooler, denser air from the water to move inward, creating circulation and wind. Most sea breezes start around midday, gain intensity as the air temperature increases, and then weaken around sunset.

One point to remember is that wind speeds are measured at a standard height of 33 feet (10 m). To get the effective midflame or wind speed near ground level, you can use reduction factors. When burning in grass fuels, reduce the 33-foot (10 m) wind speed by 50%; in shrub or brushy areas, reduce it by 60%; and in the forest understory, reduce it by 70% (Bidwell et al. 2006). These reduction factors can assist you in making a decision about whether or not a day will be suitable for your burn. If the maximum forecasted wind speed for the day is 15 mph (6.7 m/sec), and you are burning a prairie area, the wind speed near ground level would be around 6 mph (2.7 m/sec), and in a forested area it would be near 4.5 mph (2 m/sec).

Having one day with the proper weather conditions for a prescribed burn may not be too difficult to attain, but what if you were planning to conduct ten to twenty burns in a season, or needed a specific wind direction to keep the smoke from causing problems? Then it may seem as if the weather never cooperates. Only a certain number of days exist each month that are suitable for burning. If you limit the burn to specific wind direction, number of acceptable days is again reduced. A study that looked at weather constraints over a five-year period when scheduling prescribed burns in Oklahoma found that each region is different and that there are fewer acceptable burn days than you might expect (Roberts, Engle, and Weir 1999). The information in Table 5-1 shows the results of this study. High winds in the prairies of the western two-thirds of the state were the main factor for unacceptable burn days. In the forested eastern part of the state, the reason for not burning was low wind speeds within the canopy. Also, the north winds became less common later in the spring. These examples all show why planning multiple burns or desiring specific wind directions for burns can make your task more difficult. It may be necessary to burn during alternative seasons to accomplish your intended goals.

Sources for Weather Information

Numerous sources for weather information and forecasts are available, but remember that they are only forecasts, and most of the times are only good for about forty-eight hours. If you are going to properly conduct prescribed burns, you should be an amateur meteorologist, because prescribed burning depends entirely on the current and forecasted weather conditions. Be sure to find several reliable sources for weather information in your area, as this will make your decision to burn more accurate. If the weather forecast is not quite right for your burn, wait for another day when conditions will be better. The hardest thing to do is to not burn on a marginal day, but it is the safest choice, because you will not be putting your crew or anybody else in harm's way.

Be aware at times you may pull up forecasts from three different weather sources and receive three totally different forecasts. This is a positive no-burn situation. There will also be times when the high temperature will not be anywhere near what was forecast, or the wind direction and speed will be totally different. Be sure to obtain current and frequent weather forecasts before, during, and after conducting burns, as well as monitor on-site conditions.

Weather forecasts from reliable television or radio stations are always a good source, but usually they are only available at certain times of the day. Also, most of these stations use the forecast information available from the National Weather Service (NWS), so pick your sources carefully.

The NWS is an excellent source of weather information, and each state has several forecasting offices and numerous recording stations. Weather information can be obtained from the NWS offices by calling them directly and receiving spot weather forecasts, by listening to broadcasts from the NWS radio frequencies, or by looking at the NWS Web site for a specific region. The Web

TABLE 5-1 Parameters for unacceptable and acceptable burning days

	Unacceptable burning days			Acceptable burning days		
Vegetation type	*Month*	*Average*	*Range*	*Limiting weather factor*	*South wind*	*North wind*
Shortgrass prairie	January	14	11–16	High winds	6	8
	February	12	10–14	High winds	4	6
	March	18	16–20	High winds	3	5
	April (1–20)	11	7–17	Low humidity	2	4
Mixed prairie	January	13	5–22	High winds	9	6
	February	11	6–14	High winds	7	7
	March	17	16–19	High winds	7	6
	April (1–20)	11	9–13	High winds	5	4
Tallgrass prairie	January	14	12–16	Low temperature	7	5
	February	9	5–13	High winds	7	7
	March	14	12–17	High winds	7	7
	April (1–20)	10	8–11	High winds	6	5
Pine forest	January	26	21–28	Low winds	4	1
	February	25	23–26	Low winds	4	2
	March	22	16–25	Low winds	8	3
	April (1–20)	14	9–17	Low winds	6	2

Source: From Roberts, Engle, and Weir (1999).
Note: The number (average and range over a five-year period) of unacceptable burning days, the primary weather parameter responsible for limiting burning, and the average number of days in which burning conditions, constrained by wind direction, are acceptable in four vegetation types of Oklahoma over the period from January 1 through April 20.

address for the NWS is www.nws.noaa.gov. From there you can type in your city and state for local weather information and local fire weather forecasts. Each regional office Web page is different, so you will have to search to find specific fire weather information. One important thing to remember is to check when the forecast was last updated. The time of the forecast is listed above the information; make sure it is current, because weather changes rapidly.

Several private weather sources are also available on the Internet, including The Weather Channel (http://www.weather.com/), Accuweather (www.accuweather.com), Weather Underground (www.wunderground.com), and Intellicast (www.intellicast.com). These sites give reliable weather forecasts by city or area.

Many of the national forest Web sites have weather links that give fire weather forecasts for that region. To locate weather information from a local national forest, go to www.fs.fed.gov, and then click on a state or forest by name. The fire weather forecast page from the fire and aviation office in Boise, Idaho, is also available at http://fire.boi.noaa.gov/. It shows a national map of all regional NWS offices and lists their current fire weather forecasts.

The main thing about prescribed fire weather is to know the weather patterns and seasonal conditions in the area where you are burning; this is very important. If you are new to an area, be sure to seek information and advice from local people. In addition, try to go on burns with experienced personnel to get comfortable with the weather conditions in a new area.

CHAPTER 6 Fire Behavior and Fuel Characteristics

For the Lord thy God is a consuming fire, even a jealous God.

—Deuteronomy 4:24

In order for a fire to occur, three elements must be present: fuel to burn, oxygen for the flame, and heat to start and continue the combustion process. This is called the fire triangle, and without one of these elements there is no fire.

When a fire occurs or a specific fuel particle burns, four primary stages of combustion are present, and the amount of fuel consumed is distinctive for each stage (adapted from National Wildfire Coordinating Group [NWCG] 1994a):

- *Pre-ignition stage:* Fuel particles are heated by radiation and convection, causing water vapor to be evaporated from fuels, and the process of pyrolysis occurs, which is the decomposition of a compound by heat.
- *Flaming stage:* Combustible gases and vapors created in the pre-ignition stage rise and mix with oxygen. Flames then occur if the gases are heated to the ignition point or they come in contact with an object hot enough to ignite them.
- *Smoldering stage:* Combustible gases are still being released, but not enough are being released and the temperatures are not adequate to maintain flaming combustion.
- *Glowing stage:* Most volatile gases have been driven out, so oxygen can come into contact with the surface of the fuel particle, causing it to glow. This stage continues until the combustible material of the fuel particle is gone or the temperature will no longer support combustion.

Prescribed fire managers need to be able to understand and define how fire behaves and reacts as a result of these processes: for example, understanding how the behavior of a fire changes when more fine fuel is added to the fire triangle; or how they can increase the amount of fuels burned in the flaming stage rather than the smoldering stage.

Numerous methods exist for describing fire and its behavior, as well as various descriptors for fuel loading and fuel characteristics that affect a fire's behavior. These descriptions use measurable methods of time, distance, and amounts of energy released. However, they do not use temperature or try to determine the temperature of the fire because the effects of fire on vegetation are best described by the fire's behavior rather than its temperature (Alexander 1982; Johnson and Miyanishi 1995).

Figure 6-1.
The movement of fire fronts in relation to the prevailing wind. A headfire moves with the wind with higher flame heights (see flames in background of photo), and a backfire moves against the wind with shorter flame heights (see flames in the foreground).

Prescribed Fire Types

Basic fire types—headfire, backfire, and flank-fire—are named for their orientation to the edge of the fire or fireline and how they are affected by the wind or slope of the land.

HEADFIRE

A fire that moves with the prevailing wind is called a headfire (Figure 6-1). Headfires have the most rapid rate of spread and longest flame length of the three fire types (NWCG 1991). They are also the most intense (Figure 6-2) and have the hottest temperatures of all the fire types (Lindenmuth and Byram 1948; Bailey and Anderson 1980). Headfires are more damaging to shrubs and trees than are the other slower-moving fire types (Fahnestock and Hare 1964) and are normally the most difficult to control and contain. There is also more risk involved when using headfires because of the greater intensity and rapid spread rates.

BACKFIRE

Any fire that moves against, or into, the prevailing wind is a backfire (Figure 6-1). Backfires have the slowest rates of spread and shortest flame lengths of the fire types (NWCG 1991). They also have the coolest temperatures (Lindenmuth and Byram 1948; Bailey and Anderson 1980) and the lowest intensity (Figure 6-2). Backfires generally do not cause as much damage to trees and shrubs as headfires do, but damage is dependent upon fuel load, weather conditions, and tree or shrub species. The relatively slow-moving nature of backfires

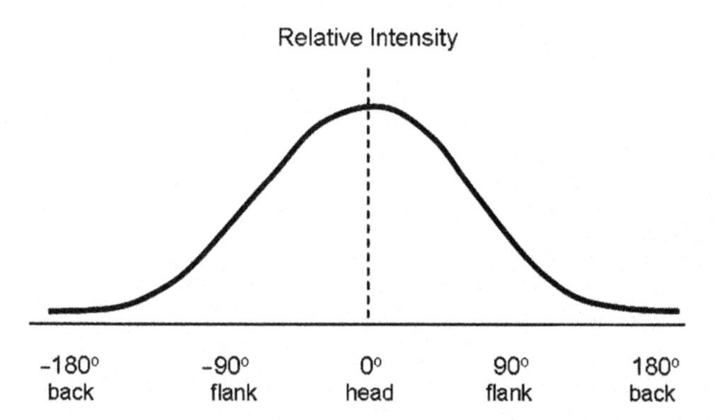

Figure 6-2.
The relative intensity of the different fire types along the perimeter of a fire (from Catchpole et al. 1982).

means that surface temperatures may be higher. Higher temperatures cause more surface organic matter to burn, which in turn can cause more damage to root and basal stem tissues (Martin 1990). Backfires are easier to contain and control than the other two fire types, but because they are slow moving, they can be more susceptible to shifts in weather.

FLANK-FIRE

A fire that moves or spreads parallel to, or at right angles (90°) to, the prevailing wind is a flank-fire. The behavior of a flank-fire is similar to that of a backfire, with temperatures and flame lengths that are normally less than those of a headfire. The intensity of a flank-fire falls between that of a headfire and of a backfire (Figure 6-2). Flank-fires can be very dangerous because their tendency is to grow outward and roll into headfires. This behavior can increase fire intensities that can damage desirable trees and shrubs; it can also cause a fire to escape the burn area or entrap personnel. Wind shifts can increase the intensity of a flank-fire and make it more dangerous by turning it into a headfire.

Flame Length and Flame Height

The flame length is the length of the flame from the lowest point where active burning is occurring to the top of the main body of flames, not including any flame flashes that arise at the tips. Flame length is measured at an angle on all wind-driven fires and can be used as an alternative observable measure of fireline intensity (Rothermel 1983) (Figure 6-1). Flame height is the perpendicular distance from the ground to the top of the main body of flames (NWCG 1994a). There also is a direct relationship between crown scorch height, flame length, ambient air temperature, and wind speed; these variables can be used to determine the lethal scorch height on trees above ground level (NWCG 1994a). Precision should not be expected when trying to measure flame length or height; measurements of 0.8 foot (0.25 m), 1.6 feet (0.5 m), 3.3 feet (1 m), 6.6 feet (2 m), and greater than 13 feet (4 m) are adequate for most purposes (Cheney and Sullivan 1997). Measurements for flame length and flame height can be recorded in inches, feet, yards, centimeters, or meters.

Figure 6-3.
Flame depth is the width of continuous burning area behind the flame front; residual flame time is the amount of time flames persist in a given place.

Flame Depth

Flame depth is the distance behind the main fire front that is covered with continuous flames (Cheney and Sullivan 1997) and is the area where active burning is taking place (Figure 6-3). It is delineated by the active advancing flaming edge at the front of the fire and the expiring edge at the back (Pyne, Anderson, and Laven 1996). Flame depth can be measured in inches, feet, yards, centimeters, or meters. The actual width of the flame depth will be dependent upon the amount, type, and condition of the burning fuel particles.

Residual Flame Time

Residual flame time, or residence time, is the period of time that active burning is occurring at a given spot or the amount of time flames persist in one place (Pyne, Anderson, and Laven 1996; Cheney and Sullivan 1997) (Figure 6-3). The length of residual flame time depends upon fuel type, fuel loading, fuel compaction, and thickness of the fuel particles. Residence time can be measured in seconds, minutes, or hours.

Fire Intensity

Fire intensity can be described as the rate at which a fire produces thermal energy (Brown and Davis 1973) or the behavior of a fire in terms of its rate of energy release (NWCG 1994a). Factors that affect fire intensity are fuel load, moisture, temperature, wind speed, plant chemistry, and topography. The three major methods used to measure or describe fire intensity are fireline intensity, heat per unit area, and reaction intensity.

The most widely used method of measurement is fireline intensity, which is sometimes called Byram's intensity (Pyne, Anderson, and Laven 1996; DeBano, Neary, and Ffolliot 1998). Fireline intensity is the amount of heat released per square foot (0.09 m^2) of fire front per second; the most important thing to remember is that fireline intensity is based on both the rate of spread and the heat per unit area of the fire (Rothermel 1983). Fireline intensity can be estimated using Byram's intensity equation (Byram 1959):

$$I = HWR$$

where I is the fireline intensity (Btu/ft or kW/m), H is the heat yield (Btu/lb or kJ/kg) of the fuel burned adjusted for moisture content, W is the amount of fuel consumed (lb/ft^2 or kg/m^2), and R is the rate of spread (ft/sec or m/sec). Fireline intensity can be reported by British thermal units (Btu) per foot per second or kilowatts per meter.

Another frequently used measurement of fire intensity is heat per unit area, the amount of heat released in a unit of area during the time that area is within the flaming fire front. Heat per unit area is normally measured in Btu per square foot or kilowatts per square meter.

Reaction intensity is another type of fire-intensity measurement. Reaction intensity is the rate of heat release per unit area of flaming fuels, or the amount of energy released per minute by a square foot (0.09 m^2) of flaming front (Rothermel 1983). The reaction intensity is measured in Btu per square foot per minute or kilowatts per square meter per minute.

Rate of Spread

The speed at which a fire moves across the landscape is a measure of that fire's rate of spread (NWCG 1994a). Factors that affect the rate of spread are wind, moisture, temperature, fuel continuity, topography, and plant community; several of these same factors also influence fire intensity (Whelan 1995; Pyne, Anderson, and Laven 1996) (Figure 6-4). Fire spread is dependent upon a percentage of the heat emitted by the flames, which is indicated by the greatest rates of spread coming from fuels that generate the tallest flames (Anderson 1968). Rate of spread can be measured in chains, miles, or kilometers per hour; feet per minute; or meters per second.

Fuel Characteristics

A number of different factors determine how the fuel will react to the fire. You will need to keep each of these in mind as you plan your burn.

Figure 6-4.
Rate of spread is how fast the fire moves over the landscape. Factors that affect rate of spread are wind, moisture, temperature, fuel continuity, topography, and plant community; several of these factors also influence fire intensity.

FUEL LOAD

Fuel load is a measure of the potential energy that could be released by a fire (Pyne, Anderson, and Laven 1996). Fuel load is the total amount of flammable fuel for the surface area of the burn unit and is normally measured on a dry weight basis. It can be measured in pounds or tons per acre or kilograms per hectare. Remember that fuel load is not the total amount of vegetation in the burn unit but what will actually burn within the unit. It is important not to confuse potential fuel (total vegetation) and available fuel (what will burn). Because these two measurements are difficult to obtain in many vegetation types, an accurate measurement of how much fuel will actually burn prior to ignition is very difficult to estimate (Whelan 1995). Even so, it is best to measure only the fuels that are expected to burn.

Management practices applied to the burn unit prior to ignition can affect the total amount of fuel available to be burned and drastically impact the objectives and goals of the prescribed fire (Figure 6-5). Land managers should follow standard management practices and allow adequate time for regrowth of fuels between fires; or if the preferred management practice increases the fuel load,

Figure 6-5.
Reduced fuel loading and fuel discontinuity can create problems with most prescribed fires and increase the amount of work required to get the fire to carry across the unit.

the land managers should allow time for some of the fuels to decompose or to be removed so the fire can be conducted safely.

Fuel architecture can impact the amount of fuel that will burn. Fuel architecture is related to the how fuel particles are arranged. For example, hardwood leaf litter that has recently fallen to the ground will burn better under the same conditions as litter that has been packed by several rainfall or snowfall events. This causes the fuels to become compacted and compressed, which makes them more difficult to ignite.

At times certain fuels will not burn because they are actively growing and holding too much moisture, thus reducing the plant's flammability and its contribution to the total fuel load. However, do not expect all green, growing plants to be fire resistant. Moisture content of fuels has usually been considered the most important factor in predicting fuel conditions and in determining fire behavior, but the chemical characteristics of the fuels are also very important in determining the flammability of plant communities (Mutch 1970). Plant fuels consist mainly of carbohydrates and lignin; depending upon the species and season of the year a variety of waxes, fats, oils, and terpenes may also be present (Philpot 1969a). These waxes, fat, oils, and terpenes have the greatest heat content of any major plant component (Philpot 1969b), so you should become familiar with the physiology and ecology of the plants in your area before determining fuel load and conducting prescribed fires.

FUEL COMPONENTS

Three types of fuel components impact the behavior of a fire (NWCG 1994a; Pyne, Anderson, and Laven 1996; DeBano, Neary, and Ffolliot 1998). The first is ground fuels, which include all combustible materials below the surface litter layer, such as partially decomposed materials like duff, punky (soft or rotted) wood, and peat layers. Ground fuels can also be living plant matter, such as roots. The second is surface fuels; they are normally the most important to prescribed burning. They consist of fuels on the surface of the ground, such as herbaceous plants (grasses and forbs), leaf and needle litter, and the upper layer of duff and dead branch debris. Third, crown fuels consist of the crowns or canopy of trees and shrubs.

FUEL CLASSIFICATION

Fuels can also be classified by their vertical orientation (DeBano, Neary, and Ffolliot 1998). Vertical fuel classification is divided into three categories: ground, aerial, and ladder fuels. Ground fuels lie on or within 6.6 feet (2 m) of the mineral soil surface. Fuels that are more than 6.6 feet (2 m) above the surface of the mineral soil are classified as aerial fuels, while ladder fuels provide a continuous fuel bed between ground and aerial fuels. Ladder fuels carry the fire from ground fuels into aerial fuels or the crowns of trees or shrubs. Ladder fuels can cause problems during prescribed burns by facilitating crown fire development, which can cause spotfires or unwanted top-kill of trees and shrubs (Figure 6-6).

FUEL CONTINUITY

Fuel continuity describes the amount of coverage, or distribution, of fuels and is the major factor affecting the spread of a fire (Cheney and Sullivan 1997). With greater fuel continuity there is a faster rate of spread, which in turn requires lower fireline intensity for the fire to spread (NWCG 1994a). Horizontal continuity is the relationship of the horizontal distances between fuel particles and relates to percentage of cover, while vertical continuity is the proximity of surface fuels to aerial fuels and affects the likelihood of a fire carrying through the vegetative canopy (NWCG 1994a). Fuel continuity is often affected by both the fire return interval and the management practices that have been implemented (DeBano, Neary, and Ffolliot 1998) (Figure 6-4).

FUEL MOISTURE

Fuel moisture is a measure of the amount of moisture in a given fuel particle and can be categorized as either live or dead fuels. Living fuels are classified as woody or herbaceous. The main disadvantage of measuring the moisture content of living plants is that the moisture conditions of these live fuels varies seasonally according to the growth patterns of the plants (DeBano, Neary, and Ffolliot 1998). Due to fine fuel loads, burn managers in Texas have been monitoring leaf moisture in both the redberry juniper (*Juniperus pinchotii*) and Ashe juniper (*J. ashei*) prior to burning so that a prescribed fire will be conducted at the best time for control of these trees (Blair, Sparks, and Franklin 2004). Burn

Figure 6-6.
Ground fuels are carried into the aerial fuels by ladder fuels. These ladder fuels include small trees and shrubs, as well as vines. Most ladder fuels can be removed by the use of periodic prescribed fires.

managers have developed a technique that allows land managers to quickly determine and monitor live juniper leaf moisture. Juniper leaf samples are obtained in the field and weighed; then samples are dried in a microwave oven at thirty-second intervals until the samples no longer lose weight. The wet weight is subtracted from the dry weight, divided by dry weight, and multiplied by 100 to obtain percentage of juniper leaf moisture. If your fine fuel loads are less than 2,000 pounds per acre (2,240 kg/ha), you will need to have a juniper leaf moisture of less than 75% to obtain adequate control of these trees when they are 4 to 8 feet (1.2 to 2.4 m) tall (Blair, Sparks, and Franklin 2004).

Figure 6-7. Fuel moisture sticks set up to monitor fuel moisture for a specific area: 1-hour (top left), 10-hour (top right), 100-hour (bottom left), and 1,000-hour (bottom right).

Dead fuels are a better measure of fuel moisture and are measured using a time lag period. The time lag period is the length of time that it takes a fuel particle to lose approximately 63% of the difference between its initial moisture content and the equilibrium moisture content (Fosberg 1970). The equilibrium moisture content is defined as the value that the actual moisture content approaches if the fuel particle is exposed to constant atmospheric conditions of temperature and relative humidity for an infinite length of time (Schroeder and Buck 1970). Dead woody fuels are grouped by size into four time lag classes: 1-hour time lag fuels, less than 1/4 inch (0.6 cm) in diameter; 10-hour time lag fuels, 1/4 to 1 inch (0.6 to 2.5 cm) in diameter; 100-hour time lag fuels, 1 to 3 inches (2.5 to 7.6 cm) in diameter; and 1,000-hour time lag fuels, 3 to 8 inches (7.6 to 20.3 cm) in diameter (NWCG 1994a) (Figure 6-7).

Following a precipitation event, 1-hour fuels such as dead grass reach the equilibrium moisture content with ambient weather within one hour. In comparison, dead woody fuels 2 inches (5 cm) in diameter are 100-hour time lag fuels because they may require up to four days at a constant temperature and relative humidity to reach equilibrium moisture content. Therefore, from a prescribed fire perspective, if the goal of your burn is to reduce fuels that are classed as 100-hour time lag fuels, then it will take longer for these fuels to dry down after a precipitation event than it will take for the smaller-size fuels to dry down. You should take this into consideration when planning and conducting burns that have large amounts of downed woody debris.

Moisture content in woody fuel samples can be measured in the field by a moisture meter called a protimeter or weighed in the field and then dried in an oven and weighed again to determine percentage of moisture content. Fuel moisture samples of woody fuels are taken from standing or aerially suspended dead trees or shrubs and must not be lying on the ground. Because of the hygroscopic nature of dead plant material, these fuels readily absorb moisture from the soil if they are lying on the ground, which will cause the fuel moisture of that fuel particle to be extremely overestimated. Herbaceous fuels are determined in the same manner, and samples must come from standing material.

The preferred moisture range of 1-hour time lag fuels for prescribed fires is 7% to 20%. When fine fuels drop below 5% moisture, spotfires are certain to happen and control of the fire is considerably more difficult. Spotfires are rare when 1-hour fuel moisture is above 11%. When 10-hour fuel moisture is between 6% and 15%, prescribed fires will spread well. This fuel size will burn up rapidly at 6% fuel moisture but will not burn at >15% (Bidwell et al. 2006).

CHAPTER 7 Fire Prescriptions

And say to the forest of the south, Hear God; Behold, I will kindle a fire in thee, and it shall devour every green tree in thee, and every dry tree: the flaming flame shall not be quenched, and all faces from the south to the north shall be burned therein.

—Ezekiel 20:47

According to *Webster's Collegiate Dictionary*, the word *prescribe* means "to lay down as a rule or directive; to establish rules, laws, or directions; to order or recommend a remedy or treatment." A prescription is a formula directing the preparation of something; in this case it refers to a fire. A fire prescription for a prescribed fire is a set of conditions under which a fire will be set to meet specific land management goals and objectives; it is also the systematically planned application of burning to meet specific management applications (Scrifres and Hamilton 1993).

Fire prescriptions are based on scientific research and experience; remember that specific recommendations must be customized for each particular burn unit. In any given case there are numerous possible fire prescriptions based on countless factors, which can vary from type of burn (reclamation or maintenance), fuel load, fuel type, fire boss experience, crew experience, and available equipment, to the management goals and objectives of that specific unit. Based on these variables, one set prescription will not necessarily work for all burn units, even when adjacent to each other.

Why do you need a prescription to conduct a prescribed burn? The first answer is safety—not only the safety of the people conducting the fire but also the safety of the people who might be impacted by smoke or by the fire itself if it were to escape. You should always have a set of guidelines, including weather conditions, under which you will or will not burn. A prescription allows you to burn when conditions are safest for everyone involved; it also makes containment of the fire easier. Second, a prescription can help you meet your management goals and objectives. The prescription will allow you to determine under what type of conditions you should burn to achieve a specific objective. This could range from burning to control brush in light fuels to conducting a fire that does not scorch or top-kill certain trees. Third, a prescription will allow for the execution of the burn in an organized and well-thought-out manner. Finally, it provides documentation of environmental conditions and the reason the burn is being conducted.

Writing a single prescription for all prescribed burns is difficult and can be limiting. Each burn unit is different and can be burned safely under a wide range of conditions. Written prescriptions can also cause problems in the courtroom, as discovered by several people who have given expert testimony in cases involving fire. When there are published prescription recommendations and a person chooses to burn outside these parameters, liability increases even if

TABLE 7-1 Potential fire prescription variation for conducting a prescribed fire

Prescription variable	*Range*
Temperature	30°F–110°F (–1°C–43°C)
Relative humidity	10%–80%
Wind speed	4*–25 mph (2–11 m/sec)
Season of burn	Winter, spring, summer, or fall

Source: Bidwell et al. (2006).
* Caution: Light winds are often variable in direction.

conditions for burning were safe. Granted, some units should only be burned under the safest prescriptions, but others can be burned safely even under extreme conditions. If written prescriptions are published for a region, then each burn unit within this area is limited to these sets of conditions. Therefore, I recommend using a wide range of weather parameters for prescriptions and modifying existing prescriptions to specific burn units as needed. Table 7-1 shows the prescription range for conducting prescribed fires. Each burn unit is different and can be burned under a wide range of conditions; the main thing to remember is to burn when it is safe.

A prime example is two burn units located right next to each other. One unit has been burned successfully for several years and is composed of a large quantity of grass and some scattered brush. This burn unit could in all probability be burned safely with low temperatures, high relative humidity, and lower wind speeds. The other unit has not been burned, nor has any other type of management practice been conducted on it for several years. The unit is covered with dense brush and a low amount of fine fuel. To burn this unit effectively will require higher temperatures, lower relative humidity, and higher winds. These two units have the same plant species and are in the same region but will require two different prescriptions for burning (Figure 7-1).

Another example of a prescription being stretched to its limit would be a burn unit located right next to safe buffer such as a lake or area of cultivation. This area could safely be burned under very extreme conditions, similar to intense wildfire situations. But if there are limits on the range of the prescription's conditions, how can such a unit be burned effectively? I know people who manage lands right next to large lakes. Many of their burn units were not managed for years, so "normal" prescribed fire is not the most practical method of management. It would also be costly and time consuming to mechanically treat these areas. But these managers understand how to use prescribed fire along with extreme weather conditions to plan an appropriate burn for their units.

Figure 7-1.
Due to the vegetation types present, the fire prescription for the burn unit on the left would be much different from the requirements for the unit on the right. Each burn unit is not only different physically but also different in terms of goals and objectives. All this information should be taken into consideration when planning burn prescriptions.

They are looking for a specific wind direction, extremely high temperatures, and low relative humidity when they burn these areas; they are essentially creating a very effective prescribed wildfire. If they were limited by a narrow prescription range, these burns would not be very successful and might not be able to be conducted at all.

A good illustration of written prescriptions limiting the amount of prescribed burning being conducted in a region is from a federal agency in Texas. This agency had a single prescription recommendation for the entire state. In the Coastal Plain region, because of the short dormant season, this statewide prescription caused problems. During the time when grass fuels are dormant, air temperatures can exceed that of the agency recommendation. These higher temperatures did not make it unsafe to burn in the Coastal Plain region, but the temperatures were not within the agency's prescription range. For the same reason, if agency personnel burned or recommended that landowners burn while these higher temperatures were occurring, the personnel were acting outside the scope of their jobs. The personnel in this region presented their problem and the facts about burning with higher temperatures to agency officials, who in turn granted this region a higher temperature range for conducting prescribed burns.

Burn prescriptions written with higher temperatures, lower relative humidity, and higher wind speeds are mainly for reclamation-type burns or fires with very low amounts of fine fuel. Under these extreme conditions a greater amount of brush control can be obtained. Fires will also have greater intensity, exhibit extreme fire behavior, and carry across lighter fine fuel loads more readily. The chance of spotfires is also greater under these prescriptions. Conducting burns on the lower end of the prescription range should only be done by an experienced fire boss and crew. Fires can also be conducted safely under these conditions if the burn unit is in an area completely surrounded by nonflammable material for a great distance, such as a lake or cultivated fields.

Prescriptions with lower temperatures, higher relative humidity, and lower wind speeds are more suitable for maintenance-type burns or when reintroducing fire into a ecosystem long excluded from fire. Burning under these conditions is also more suitable for increased safety and burning in unfamiliar areas or unfamiliar fuel types, as well as when burning with an inexperienced crewv. These conditions are well suited for creating mosaic-type burns for wildlife habitat. Many times the same effects of burning under an extreme prescription can be achieved with safer conditions by using a cooler prescription with greater amounts of fine fuel.

Rules for Burning

Use the "60:40" rule for safety when conducting burns. The "60:40" rule can be stated two ways. First, burn when both the temperature and relative humidity are between 60 and 40 (Wright and Bailey 1982; McPherson et al. 1986). Second, burn when the temperature is less than 60°F (16°C) and the relative humidity is greater than 40% (Bidwell et al. 2006). Keeping this rule in mind will make burning safer and easier on most burn units.

Another important rule to remember is to never burn within twelve hours of predicted wind shift or frontal passage because of the problems you may encounter. The main reason not to burn before a predicted wind shift is that it is only a prediction, not a guarantee, of the time of day that the wind will shift. We all know that weather forecasts are not always reliable; would you stake your reputation or someone else's life on a forecast? How many times have you been caught without a jacket early in the day because a front was forecast to come through in the late afternoon but came through around lunchtime? The front arrived several hours sooner than forecast. If you try to conduct a prescribed burn before a wind shift, the same thing is likely to happen. You could be putting in the backfire when the wind suddenly shifts, and your backfire becomes a headfire racing to the other end of the burn unit; more than likely the firebreaks around the burn unit will not stop this newly formed headfire (Figure 7-2).

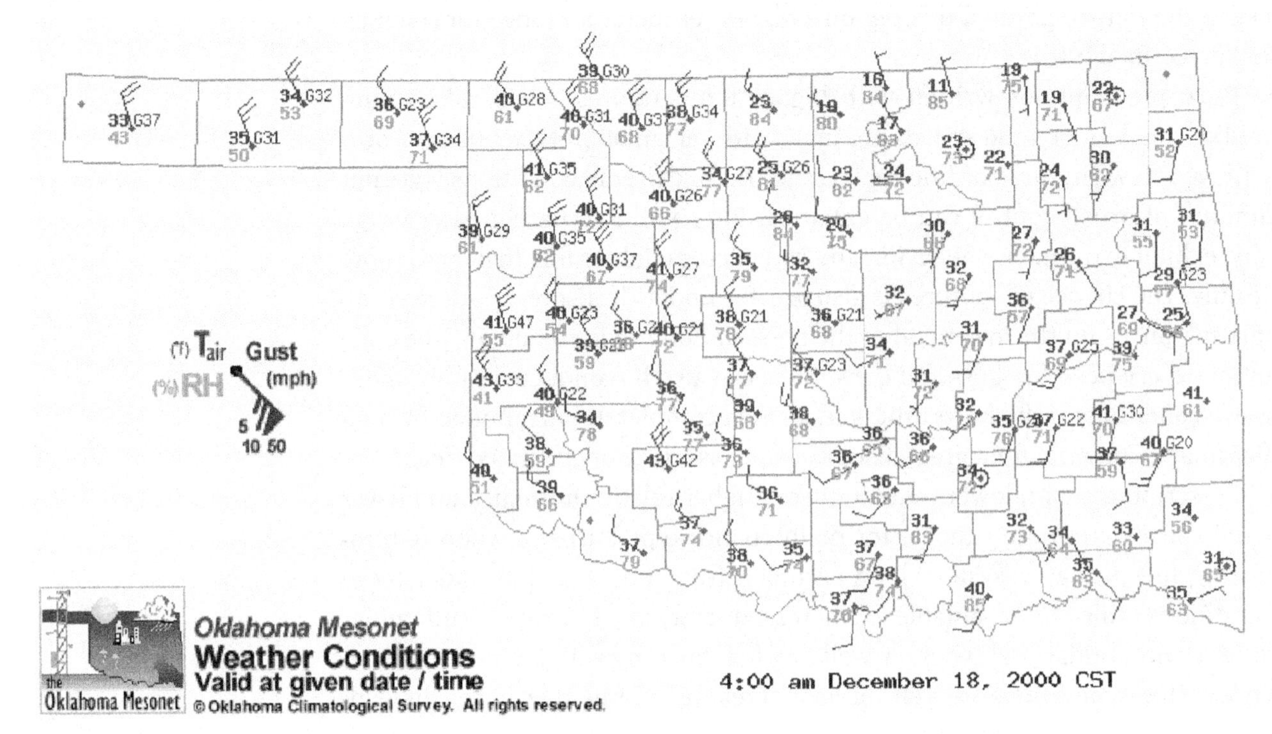

Figure 7-2.
The wind shift line on this frontal boundary is well defined. The winds behind the front are gusting strongly; east of the line, the winds appear perfect for burning. How much time do you have before the winds switch? Do not burn within twelve hours of a predicted wind shift (from the Oklahoma Mesonet).

Probably one of the worst examples of this situation occurred in north-central Oklahoma in the early 1990s. A landowner was going to burn his pasture on the north side of a major four-lane divided highway. The two people conducting the burn were inexperienced with minimal equipment. This actually should not even be called a prescribed burn because there was nothing planned about it. The wind was out of the south when they lit the north end of the area to be burned. Just after they lit the north side, the wind shifted and came out of the north; they had forgotten to check the weather forecast for the day. If they had, they would have known a cold front was approaching with strong northerly winds. The fire moved south and entered an area next to the highway where a large number of round hay bales were stored, catching them on fire. The smoke covered, and eventually shut down traffic on the highway, but not until one person had been killed and a dozen vehicles had collided. The potential loss of life and property, as well as having to live with the consequences, are not worth the cost of burning within twelve hours of a predicted wind shift or frontal passage.

One rule that people usually learn the hard way is to only set fires that they believe they can extinguish if the fires escape. If you feel the conditions are too extreme and suppression would be exceptionally difficult, wait for a better day. It is always better to err on the side of caution. We figured this out one day in eastern New Mexico while burning research plots. We got everything ready and lit one small plot. The fire immediately burned across that plot, jumped a dozed line 8 feet (2.4 m) wide, burned through another plot, and then for-

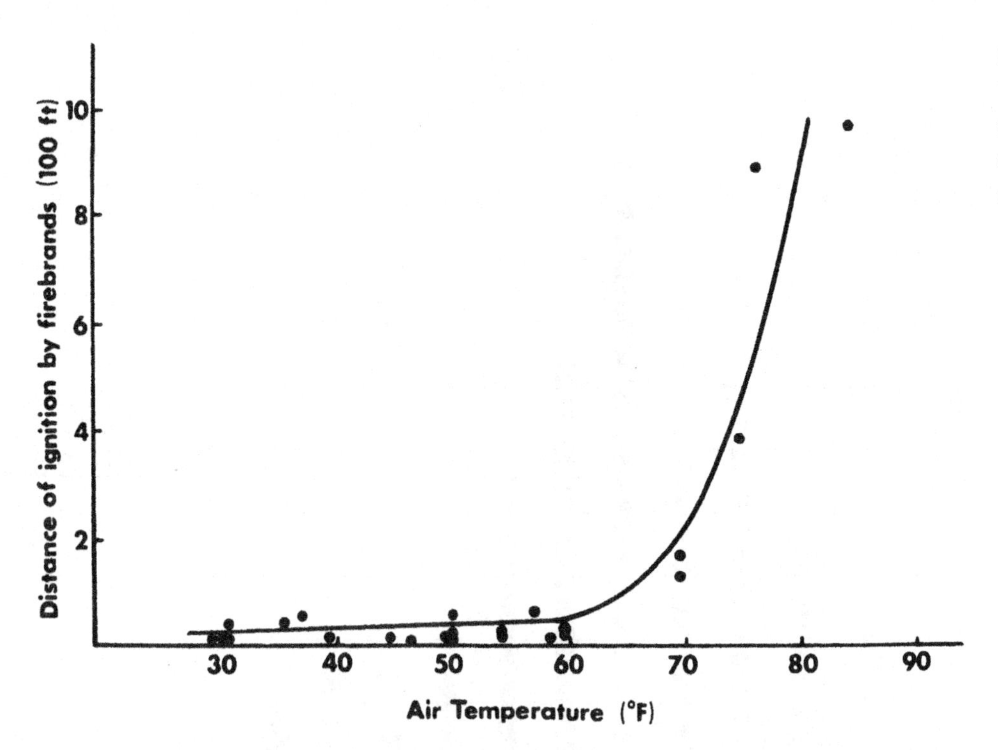

Figure 7-3.
The graph shows the distance spotfires can occur downwind at specific temperatures (Bunting and Wright 1974).

tunately was stopped by the 16-foot (4.8 m) dozed firebreak on the perimeter. This fire moved so rapidly there would have been no way to stop or contain it if it had escaped.

Spotfire Prevention

Spotfires have always been, and will always be, a problem that prescribed fire personnel have to contend with on prescribed burns. Spotfires can cause mental and physical stress on personnel and can cause injury or even loss of life, as well as monetary damages and loss of public or official support for the prescribed burning program. From interviewing many private and public land managers in Oklahoma, I have learned that spotfires, or the risk of escape (liability), are the main reason many of them do not conduct prescribed fires. This is likely to hold true in other areas as well.

Temperature, wind speed, and relative humidity are the most important weather factors that personnel can use to predict and monitor prescribed fire behavior. The danger from firebrands—burning embers produced from the burn that travel in the air and land outside of the burn unit—is lower if the ambient air temperature is below 60°F (15°C) when conducting prescribed burns (Bunting and Wright 1974) (Figure 7-3). This study proved that the air temperature helps determine the distance downwind that spotfires may occur in relation to the main fire. It also demonstrates that a spotfire can occur at any given temperature within the study range. The study shows that air temperature is not a great predictor of whether or not a spotfire is going to happen.

Wind speed could also be considered as a potential source of spotfires. Wind

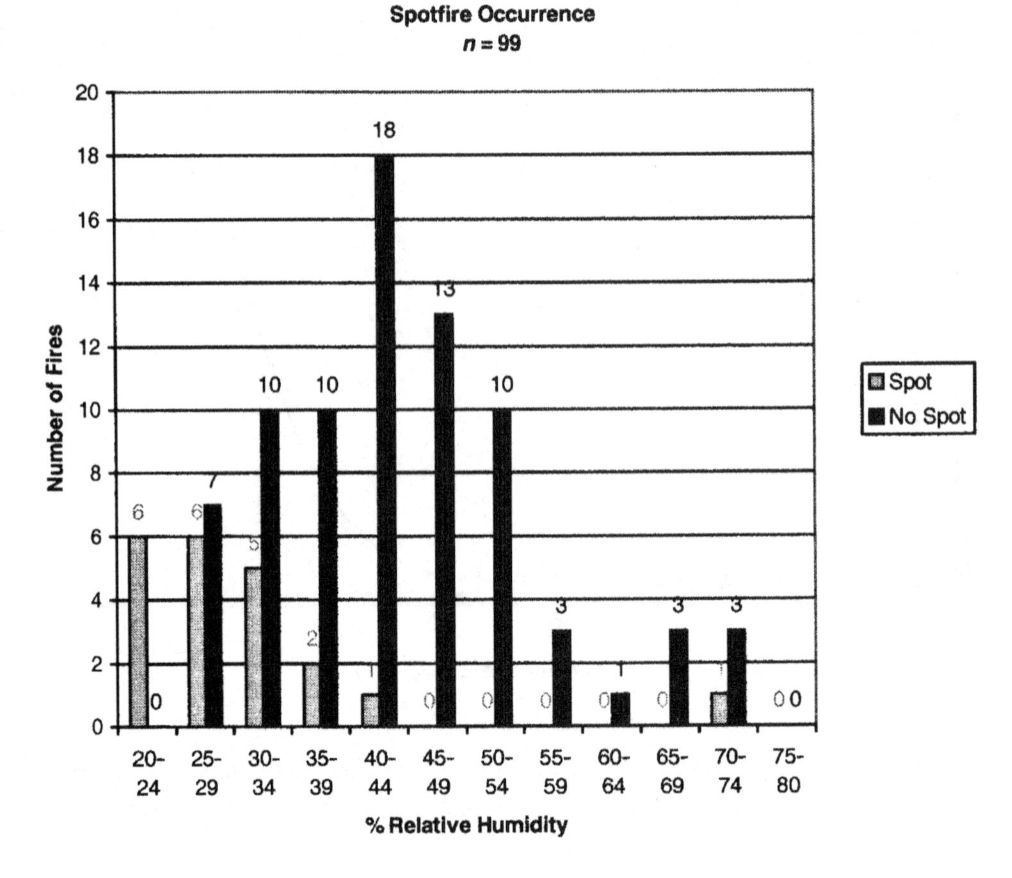

Figure 7-4.
Of the ninety-nine prescribed fires conducted throughout Oklahoma from 1996 to 2002, twenty-one burns had spotfires, but only two spotfires occurred when relative humidity was greater than 40% (Weir 2007).

speeds of 8 mph (3.6 m/sec) are needed to ignite and burn standing fuels (Britton and Wright 1971); but with winds over 20 mph (8.9 m/sec) firebrands and other debris become problems (Wright and Bailey 1982). Most prescriptions do not recommend burning with winds greater than 15 mph (6.7 m/sec) unless special circumstances allow it, so wind speed is not a good forecaster of whether or not a spotfire will occur.

The final weather factor that should be considered is relative humidity. Earlier work has shown that below 40% relative humidity, fine fuels ignite and burn easily; above 40% relative humidity, ignition slows, rates of spread are significantly reduced, and danger from firebrands is not as significant (Britton and Wright 1971; Lindenmuth and Davis 1973; Green 1977). This 40% relative humidity threshold goes back to the "60:40" rule for safe burning.

As indicated from the research, several weather-related thresholds influence the occurrence of spotfires. Narrowing down the causes of spotfires to one key weather factor will help fire personnel focus on a single variable, possibly reducing the likelihood of spotfires. In a study that reviewed weather data from ninety-nine prescribed burns conducted throughout Oklahoma, one weather variable stood out as the main contributing factor of spotfire incidence—low relative humidity (Weir 2007) (Figure 7-4).

This research validates that the threshold value of 40% relative humidity is

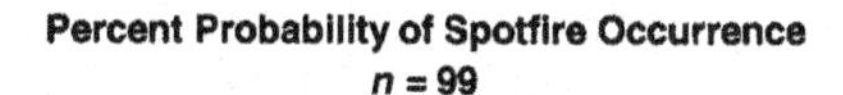

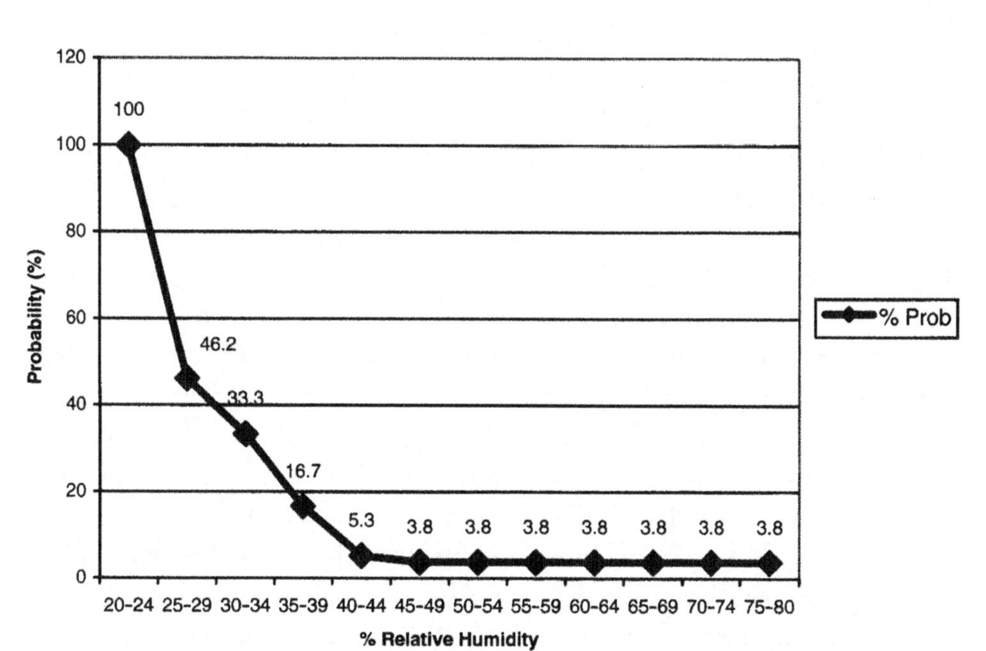

Figure 7-5.
The probability of a spotfire occurring below 40% relative humidity increases as the relative humidity decreases (Weir 2007).

an excellent rule of thumb to follow when conducting prescribed burns. This does not mean that no prescribed burns should be conducted below 40% relative humidity. Many regions of the United States, as well as specific burn units, require low relative humidity to meet goals and objectives of that given burn. But the personnel should keep in mind that there is an excellent probability for a spotfire to occur if the burn is conducted below 40% relative humidity. In the same vein, crew members should be ready for spotfires on all prescribed burns no matter what the relative humidity. For example, one spotfire was recorded in the study when the relative humidity was 73% (Weir 2007) (Figure 7-4), which demonstrates that spotfires can occur at relative humidity values greater than 40%.

Because most spotfires occur at or below 40% relative humidity, what is the probability of a spotfire occurring at the 40% threshold value or any other relative humidity value? This information would be important for personnel to know as they prepare for and conduct prescribed burns to help reduce the risk to personnel and property. A set of probabilities for a spotfire occurring at certain relative humidity values was developed from this study (Weir 2007) (Figure 7-5). It reports the probability of a spotfire occurring on any prescribed burn when the relative humidity is between 20% and 80% was determined to be 21%, or approximately one out of five burns. There was also a 41% probability for a spotfire occurring when the relative humidity was below 40%, and a less than 4% probability when the relative humidity was above 40%. This is a significant difference, and the 40% threshold should be considered when inexperienced personnel are conducting prescribed burns, when heavy fuel loads are adjacent to the burn unit, or if an escaped fire could bring possible public scrutiny or litigation. Again, this is not condemning prescribed burning below 40% relative humidity but bringing attention to the problems that can result from burning outside the 40% parameter.

If burns are conducted below 40% relative humidity, does the probability of spotfires change at lower relative humidity values? As Figure 7-5 shows, there is a difference in probability at each 5% drop in relative humidity below 40%. There also appears to be another threshold at less than 25% relative humidity. At this point the probability of a spotfire occurring is 100%. Below 25% relative humidity crew members should be prepared for a spotfire. The spotfire probability drops to 46% in the 25% to 29% relative humidity range, which reduces spotfire risk over half with an increase in relative humidity of just a couple of percentage points. When the relative humidity is between 30% and 34%, only one out of three prescribed burns is likely to have a spotfire. So even within the range of 20% to 40% relative humidity, there is a large difference in the probability of a spotfire occurring.

With this information, personnel can determine spotfire potential when considering burn units or burn days. The information can also assist burn managers when considering crew size and equipment needed, as well as possibly reduce anxiety when burning below 40% relative humidity. Most of all, inexperienced personnel can use this to help reduce risk (liability) and increase safety for crew members. The main item to remember is to burn when conditions are safest for the crew and surrounding neighbors.

Each region and vegetation type within the United States, as well as individual burn unit, is different, and each one will require a different fire prescription. Describing each of these prescriptions is a separate book within itself. For local or regional assistance with prescriptions, contact your local county extension agent, NRCS, state wildlife or forestry department, university, or private consultant for information on weather conditions to use when conducting prescribed fires in your area.

CHAPTER 8 Fire Plans

In flaming fire taking vengeance on them that know not God, and that obey not the gospel of our Lord Jesus Christ.

—II Thessalonians 1:8

Prescribed fire plans and fire policy go hand in hand because the *plan* is affected by how the *policy* is written. The policy should determine what you need to have in order to properly conduct prescribed fires, and the plan should determine the specific requirements for each individual burn unit. Sometimes the fire plan may require more detail than the policy dictates for a burn to be conducted safely, but the plan should never have less information than required by the policy.

Prescribed burning is no different from any other management practice in the terms of needing a written plan to follow. The most important reason for having a fire plan is to plan out the actions of the fire boss and crew. This will help you determine how, what, when, and where the burn should be conducted. Fire plans assist in defining the burn unit boundaries, along with determining what type of ignition technique is the safest and easiest to use. Most important, a fire plan helps you determine the safest and easiest way to complete your tasks before, during, and after the burn.

The second reason is to set the prescription or range of weather conditions under which the burn can be conducted. The plan should include ranges for temperature, relative humidity, and wind speed and direction and could include 1-, 10-, and 100-hour fuel moistures as well. The most important thing to remember is not to limit the plan by creating too narrow a range, or a range that is difficult to attain.

The third reason to have a fire plan is for legal purposes. A few states require some type of fire plan to be submitted for each burn as part of the state burning laws. For example, under Florida's prescribed burning law a fire plan must be submitted before a burn can begin; the law also dictates specifically what should be included in the plan (Brenner and Wade 1992). Be sure to check the laws in the state where you are burning to make sure you are following the applicable burning statutes. This may save you from being fined or protect you from liability, as in the case of states with a prescribed burning act.

Your employer may require you to file a fire plan before you begin your burn; completing a fire plan can protect the employee from personnel liability and possible litigation. If employees violate written policy, they become personally liable for damages because they are acting outside the scope of their employment; employees could also lose their jobs if they violate the policy.

Fire plans will also help the fire boss consider any social impacts of the burn, including smoke management concerns and traffic patterns or problems, along with other public health and safety issues. These may not be big concerns, but in areas experiencing growth and the problems associated with urban sprawl, it can be a major problem and should be an important part of each fire plan.

What should be included in a fire plan? This is a good question to which there is no specific answer. Each employer, organization, burn unit, or region is different and will require varying amounts of information (NWCG 1986; Wade and Lunsford 1989; Bidwell, Weir, et al. 2003). The main thing to remember is not to be so detail oriented that you limit yourself; on the other hand, do not be too brief, because having too little information can also cause problems, especially if there are legal ramifications. Also, if you have to comply with state laws or employer requirements, be sure to have all the information required by them on each plan. The following is a list of information that would be appropriate to include in most fire plans:

- *Description of burn unit:* This should be both physical and legal. The legal description would include the portion of section or sections, range, and township. For instance, the burn unit may use the following legal description: "the NW/4 of section 6, T18N, R1E." The physical description would be how the unit looks and could include a description of the soils, vegetation, topography, fuel load, and any other geographical features that would be relevant to the burn.
- *Location description or how to get to the unit:* This will allow you to communicate to others exactly where the burn unit is located and how to get to it: for example, "From the junction of SH 3 and SH 334 go 5 miles west on SH 3; then go 2 miles north on County Road G; turn east into unit. This will be the southwest corner of unit."
- *A list of the burn objectives:* The list shows the goals and objectives to be accomplished by the burn. Some examples include burning for eastern redcedar control, brush suppression, livestock forage, or wildlife habitat improvement.
- *Maps and aerial photographs of the burn unit:* These maps and photos should show as much detail as possible. Be sure to outline the burn unit and mark any hazards or problem areas. It is also good to include an area map or aerial photograph to show where the burn unit is in relation to the surrounding area.
- *Prescription parameters:* Provide a range of weather conditions under which the burn should be conducted. This can include temperature, relative humidity, wind speed and direction, fuel moisture, and fire behavior. Remember not to limit yourself by making the ranges too narrow.
- *Observed weather conditions:* Leave areas blank for recording on-site weather conditions the day of the burn, such as temperature, relative

4.0 WEATHER CONDITIONS FOR BURNING IN GRASS PASTURES
(Bidwell et al. 2004, Wright and Bailey 1982, Launchbaugh and Owensby 1978)

4.1 TEMPERATURE (°F)
1345 1445 1600
DESIRED 40°–70° ACCEPTABLE 30°–110°
OBSERVED 45 47 47 JW

4.2 RELATIVE HUMIDITY (%)
DESIRED 35–50 ACCEPTABLE 10–80
OBSERVED 45% 43% 46% JW

4.3 WIND SPEED (MPH)
DESIRED 5–10 ACCEPTABLE 4–25
OBSERVED 5 5 5 JW

4.4 WIND DIRECTION
PRESCRIBED **Any Northerly direction**
OBSERVED NE NE NE JW

Figure 8-1.
In your fire plan, include a place for observed weather conditions before, during, and after the burn; also, document the time and person observing the weather.

humidity, wind speed and direction, along with time and person observing the weather (Figure 8-1).

- *Firebreak types:* List the types of firebreaks that are used along the boundary of the unit. For instance, "on the east and north—county road; west—creek; south—mowed and disked."
- *Ignition plans, both maps and written:* This will help plan out the ignition pattern and help the fire boss foresee any problems or hazardous areas that may need to be addressed (Figure 8-2).
- *Smoke management plan:* This plan includes a list of potential downwind smoke problems and how to manage them, as well as smoke trajectory maps and other smoke management plans required by agency or state law.
- *Preburn notification:* Keep a list of contacts who need to be notified in the months or days before burn. In Oklahoma, for example, people conducting prescribed burns who want to have limited liability must contact adjoining landowners within sixty days of burning, as well as within twenty-four hours of igniting the fire. It is good policy to write down on the plan when you contacted people and whom you contacted in these preceding days.
- *Burn day notification:* Keep a list of contacts, including phone numbers, who should be notified on the day of the burn, such as supervisors, neighbors, fire departments, sheriffs, division of forestry representative, and so on. When contacting these people, be sure to write down time contacted and with whom you spoke. Writing down the names and times can help you avoid potential problems. While conducting a burn in north-central Oklahoma, after contacting all concerned parties, the assistant chief of a local fire department showed up and started

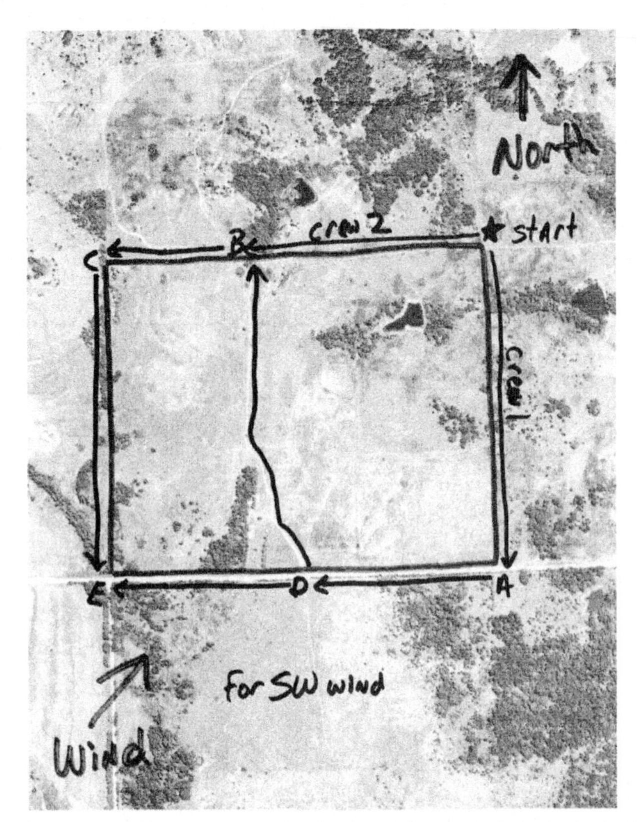

Figure 8-2. This is an example of an ignition map used to plan the ignition pattern for a burn unit. It is also important to use a written ignition plan in conjunction with the map. The following is an example of a written ignition plan for this burn unit: For a southwest wind, begin in the northeast corner of the unit. Crew 1 will go south along the east line, igniting backfire until they reach point A, at which time they will stop. Crew 2 will proceed west along the north line, igniting backfire until they reach point B, at which time they will stop and notify Crew 1. Once Crew 2 reaches point B, and there is an adequate amount of blackline along the north and east sides, Crew 1 will proceed to point D and stop. At point D, one person with a torch will follow the lease road, lighting on the east side of it north to point B. Once the torch bearer has reached point B, Crew 2 will proceed to point C and stop. When there is adequate blackline along the north side from point B to C, then Crew 2 may proceed south to point E; at the same time Crew 1 may also proceed to point E.

complaining that I had not called the fire in. I pulled out my fire plan on which I had written the name of the dispatcher and time the local fire department had been contacted and showed it to the irritated assistant chief. His attitude immediately changed, and he got in his vehicle and left.

- *Equipment needed or present:* Have a checklist of equipment that could be present or have blanks to list the equipment that is present on burn day. Again, do not limit yourself by listing more equipment than is needed or equipment that may not be available because it has mechanical problems or is being used for other duties.
- *Ignition hazards:* This should include hazards within and around the burn unit. It should also have plans for how to deal with and overcome

Figure 8-3.
Be sure to go over the ignition plan and ignition hazards with the crew prior to burning; this is very important for the safety of the crew and people outside the burn unit (photo courtesy of Alaina Thomas).

these problems. Ignition hazards could be slopes, ditches, buildings, utilities, heavy fuels, volatile fuels, or brush piles next to the fireline (Figure 8-3).

- *Escaped fire plan:* This must state what will be done if the fire escapes, including what are you going to do and how you will handle the situation. For example: "In the event of an escape, all ignitions will cease until the escaped fire is suppressed; standard firefighting methods will be used to suppress an escaped fire." The escaped fire plan should also include what to do if the escaped fire cannot be contained quickly, including when to call and who will call for assistance from local fire departments or other agency personnel. It should also contain plans to use areas around the burn unit that could help stop the fire or be used as places to set backfires, such as roads, creeks, or cultivated fields.
- *Safety concerns:* These concerns could be areas that are ignition hazards, as well as other areas that could cause safety problems for personnel or equipment. Be sure to point out these areas and notify the fire crew prior to ignition. Safety concerns may possibly include areas of heavy fuel, intense smoke, wet areas that can cause vehicles to become stuck, ditches, or parts of the fireline that are impassable to vehicles (Figure 8-4). Make sure to have plans for how to overcome these obstacles, and make sure the crew knows where these problems are located and how to handle each of these areas of concern.
- *Mop-up plans:* This should explain how the mop-up will be conducted,

Figure 8-4.
Make sure all safety concerns are listed and personnel are notified. Wet areas can cause problems for vehicles, personnel, and the safety of the burn in general. If a vehicle becomes stuck, it cannot be used until it is recovered. This takes time, equipment, and people away from the burn.

and to what level the unit will be mopped up. It may also list how many personnel are needed, as well as the period of time those personnel will monitor the unit after the burn.

- *Pre- and postburn management:* It may be necessary to have pre- and/or postburn management strategies to manipulate fuels, control grazing, change timber harvest practices, or manage for specific wildlife species.
- *Signatures and date:* Have a place to put the date the plan was prepared and signed by the person completing it. You may also be required to have the approval of supervisors, state agencies, or local authorities on each burn plan, so include places for their signatures.
- *Go/no-go list:* This list gives the fire boss a checklist to go over on the day of the burn. A go/no-go list may include questions such as, "Are all neighbors and fire departments notified?" "Are required personnel and equipment on-site?" "Are the weather conditions within prescription?" If any question is answered "no," then you should not burn.
- *Postburn personnel questionnaire:* This questionnaire could assist with safety issues and equipment problems that occur on the burn. Some possible questions could be, "Did you understand the burn plan?" "Were instructions given clearly?" "Did you ever feel unsafe at anytime? If so, when and why?" "Did you ever have any problems with the equipment? If so, list equipment type, number, and problem." "Is there anything that could be done to make the next burn safer, or easier?"
- *Postburn evaluation:* This evaluation will assist in determining if goals and objectives of the burn were met. It looks at percentage of brush control, scorch height, or other fire effect measurements.

Remember that there is no perfect prescribed fire plan; fire plans are as different as the burn units they are describing. Each fire plan may require different information or planning; one burn plan may need more information about a specific topic than other plans. Burn plans should be modified to meet local needs, as well as adapted to the region. Also, some fire plans are better prepared than others, with more planning and preparation being emphasized. The more fire plans you write, the easier it will become to write good ones. The most important item to remember when preparing a plan is not to limit yourself or crew by being too specific. Be sure to include all of the necessary information, but do not clutter a plan with pointless information that could harm your burning program if you do not adhere to the plan.

CHAPTER 9 Personal Safety

When thou passeth through the waters, I will be with thee; and through the rivers, they shall not overflow thee: when thou walkest through the fire, thou shalt not be burned; neither shall the flame kindle upon thee.

—Isaiah 43:2

Usually, conducting prescribed fires under most circumstances is not an extremely taxing physical activity, especially when everything is working well. But when problems arise or when you work on a challenging unit, burning can be both mentally and physically demanding work. These challenges can test the fitness of every crew member and serve as a reminder of why you need to keep personal safety issues at the forefront of all burn activities.

Health Hazards

Heat stress can be one of the leading health-related problems encountered while conducting prescribed burns; it can be particularly prevalent during an escaped fire or when burning during warmer months. It is very important for everyone on the burn crew to understand how to avoid and treat heat-related problems.

Heat stress occurs when the body's temperature rises above safe limits and when the water lost by sweating is not replaced. If fluids are not adequately replenished, the body's temperature-regulating processes begin to break down and body temperature increases. This causes people to have difficulty completing tasks and compromises heart and circulatory functions.

Heat-related injuries can take place when both air temperature and relative humidity are high and when there is a lack of wind or air movement. They can also occur when people are conducting strenuous tasks, working around areas of intense radiant heat, or are wearing heavy or nonbreathable protective clothing.

HEAT CRAMPS

The first level of heat stress is heat cramps, which are painful involuntary muscle contractions caused by a failure to replace fluids and electrolytes, such as sodium and potassium (Sharkey 1999). Though not life threatening, if they are not treated, heat cramps can lead to other heat-related problems. Heat cramps can be treated by stretching the affected muscles and by replacing fluids and electrolytes. When you are in the field, electrolytes can be replaced with tomato juice, sport drinks, and lightly salted water. Heat cramps are less likely to occur with adequate fluid intake and a diet that includes bananas, oranges, fresh salads, and a light sprinkling of table salt with meals (Sharkey 1997a).

HEAT EXHAUSTION

The next level of heat stress is heat exhaustion. Symptoms are weakness or extreme fatigue; staggering walk; wet, clammy skin; headache; nausea; and collapse (Sharkey 1997a). Heat exhaustion is caused by insufficient fluid intake or loss of salts from the body, or a combination of both. The fluid loss causes a reduction in blood volume that drastically inhibits work due to lowered blood oxygen levels; the loss of salts reduces the working capacity of muscles. Treatment for heat exhaustion includes replacing fluids and electrolytes, along with resting in a cool environment. In the field, such an environment would be a shaded area where a breeze is blowing, a shaded area where water could be misted over the victim, or a vehicle with the air conditioning running.

HEAT STROKE

The final heat-related disorder is heat stroke. This is a medical emergency caused by failure of the body's heat controls. Sweating stops, and the body temperature rises dangerously high. Symptoms are hot, often dry skin; body temperature above 105.8°F (41°C); confusion; incoherent speech; delirium; loss of consciousness; convulsions; or coma (Sharkey 1997a). If any of these symptoms occurs, seek medical attention immediately. While waiting for assistance, start cooling the victim rapidly with cold water or ice and fan the person to increase evaporation.

PREVENTION

The main preventive treatment for these heat disorders is to replace lost fluids. In order to stay well hydrated, you should drink water or juice before, during, and after the burn is over. When performing difficult tasks in a warm environment, replace fluids at a rate of 1 quart (0.95 L) per hour. Be sure to drink water, juices, or sports drinks, and limit the intake of caffeine and alcohol (Sharkey 1997a). Be sure to start drinking before you become thirsty or quit sweating.

You can use several methods to help prevent heat-stress problems while conducting burns. The first is to wear loose-fitting clothing that will permit air movement through the fibers. The clothing should also allow for the evaporation of sweat and should not insulate you but permit the transfer of heat. Remember that clothing should be nonflammable and made of natural fibers.

Other ways to avoid heat disorders while conducting prescribed burns are to become acclimated to the heat and to maintain or improve your aerobic fitness level. If you do not work outdoors very often, you should try to acclimate to the outside temperature prior to burning activities. It takes five to ten days of heat exposure to initiate sweating at a lower temperature; increase sweat production; improve blood distribution; and decrease heart rate and skin and body temperatures (Sharkey 1997a). Acclimate by conducting outdoor activities and slowly increasing the amount of time you are working in the heat. It is also advisable to periodically exercise in hot environments to assist with the

Figure 9-1. To avoid heat-related problems on the fireline, make sure that personnel pace themselves or have them periodically change tasks (photo by Stephen Winter).

acclimation process. The better physical shape you are in, the less risk you will have of succumbing to heat stress. If possible, try to maintain or increase your aerobic fitness through exercise prior to the prescribed fire season.

To avoid heat-related problems on the fireline, make sure that personnel pace themselves at whatever task they are conducting. It may also be necessary to have personnel periodically switch jobs or equipment. If the task is physically demanding or requires working next to extreme heat, be sure to change out personnel at regular intervals to prevent heat stress (Figure 9-1).

We learned this lesson one summer on our first burn of the season. The temperature was 107°F (41.6°C) in the shade, and we were burning ungrazed tallgrass prairie. We had a new person helping us conduct the burn; he was in top physical shape and acclimated to working outdoors, so I had him igniting the fire. He lit about 100 yards (91 m) of fire; he then turned to me and asked me to take the drip torch. I thought he did not think he was doing an adequate job, but I told him he was doing fine. He then looked at me with fatigued eyes and said, "No, you need to take the torch." I could tell then that he was having problems with the heat. I only ran the torch for a short distance before I also felt the same way. The heat from the fire was extreme, and you could not escape it because the air temperature was so high. I had another person take over my position. On that burn we learned that the drip torch operator had the hottest task on the entire fire. We changed ignition personnel every 100 yards or so. During different times of the year, different tasks are more susceptible to heat-related problems than others, so make adjustments accordingly.

Avoid working next to a heat source if at all possible. Let the fire burn down, or move away before performing your required task. So many times I see people trying to mop up a site or even operate a drip torch right next to an area that has just been lit. The flames and heat are so intense that the crew cannot safely get next to the fire. Instead, wait for the flames and heat to die down so the area can be taken care of properly with the least amount of danger to personnel. If possible, do the hardest and hottest work at night; if that is not possible, work during the cooler morning or evening periods. This could mean setting backfires or burning in heavy fuel loads when fire behavior conditions are more favorable for personnel working right next to the fire.

I have conducted numerous burns when the air temperature was greater than 100°F (37.7°C) and have not had any major heat-related problems with the crew. The main reason is that we keep an eye out for each other and know which tasks are more prone to cause heat stress. Rotate personnel frequently, and make sure plenty of fluids are available on-site to keep everyone hydrated. If you have personnel who might be overcome by heat problems, place them in areas or tasks that will not put them at risk, such as driving vehicles or in lookout positions. There may also be times when you will have to make a person sit down who otherwise will not stop working. It is great to have personnel who will work hard at their tasks, but under extreme conditions personnel need to take a break to avoid becoming overheated. You want them to be able to resume working when they are needed again, instead of being unavailable because they have become sick.

HYPOTHERMIA

Heat is not the only concern of people conducting prescribed burns; hypothermia can also cause problems during the winter and early spring. I have spent more nights shivering while patrolling fire lines than I care to remember. Working next to the heat of the fire all day increases your body temperature and saturates your clothing with sweat. You can then become rapidly chilled and are a prime candidate for hypothermia once the cool evening temperatures kick in. If you have to mop up or patrol the fireline at night, be sure to bring along a jacket or dry clothing to change into to make your stay safe and comfortable.

SMOKE

Another safety hazard that personnel conducting prescribed burns have to contend with is smoke. The main inhalation hazard from the smoke is carbon monoxide (CO), which at low levels can cause headaches, nausea, impaired judgment, and slowed reaction times (Ward, Rothman, and Strickland 1989). At higher levels, CO may increase the risk of cardiovascular disease (Sammons and Coleman 1974; Sharkey 1991). Another hazard is aldehydes, mainly in the form of formaldehyde (HCHO), which are irritants that can cause a loss in olfactory sensation or sore throats; they are also known carcinogens (Partanen 1993). Respirable particulate matter (<10µm), total particulate matter, and crystalline silica are other smoke-inhalation hazards. These particles can cause sore eyes, cough, runny nose, sore throat, and an increase in upper respiratory problems; long-term exposure could cause thickening of lung tissue, which will limit pulmonary ability (Sharkey 1997b; Reinhardt and Ottmar 1997b, 2000).

Smoke exposure has been found to be more of a problem when working on prescribed burns than on wildfires. Researchers on this study think that on prescribed burns personnel believe they have to control the fire no matter what and thus place themselves in areas with greater concentrations of smoke. According to the same study, there appear to be differences in the amount of smoke exposure depending on which task the person is handling on a prescribed burn (Reinhardt and Ottmar 1997a). The direct attack or suppression of a spotfire and holding or patrolling the fireline are tasks that have a higher smoke exposure risk than others. Mop-up and ignition duties have less exposure to smoke hazards.

The health-related effects of smoke exposure can cause short-term and intermediate-term health problems. No data exist and little is known about the long-term effects of smoke exposure, but people have used cigarette smoking, urban air pollution, and structural firefighting for models to determine possible risks (Sharkey 1997b). Studies on wildland firefighters have shown a decline in lung function from the beginning to the end of a work shift, and from the beginning to the end of the fire season. The firefighters show a return to normal respiratory values after a period in which they are free from exposure to smoke; however, little is known of the long-term effects of smoke inhalation (Betchley et al. 1997).

Personal Safety

Personal safety should be the primary goal of every person on a prescribed fire. We have to watch out for the safety of both ourselves and others because prescribed fire is no different from any other activity in that it has its share of accidents, injuries, and even fatalities. We should work to learn from our mistakes and correct them so the next burn will be safer than the last. The USDA–Forest Service has developed a set of safety guidelines over the years. These guidelines are known as the ten standard fire orders and the eighteen situations that shout "watch out." They are great guidelines for wildland firefighters, but prescribed burning is not firefighting; it is fire setting. Because these situations are somewhat different, I have adapted the wildland fire orders to fit prescribed burning and labeled them "Prescribed Fire Orders" and the "Prescribed Fire Situations That Shout 'Watch Out'":

PRESCRIBED FIRE ORDERS

- Read over the burn plan, and go over the burn unit before you ever begin igniting fire.
- Extinguish all smoldering objects along the fireline after the burn.
- Set fires as quickly as possible, but remember that safety comes first.
- Initiate all actions based on current and expected fire behavior.
- Recognize current weather forecasts and conditions, and obtain on-site weather often.
- Ensure that crew members understand the given instructions.
- Obtain current information on the prescribed fire status during the burn.
- Remain in communication with crew members.
- Determine proper ignition technique and deployment of both crew members and equipment for each burn.
- Establish lookouts in areas where spotfires are likely to occur.
- Retain control at all times.
- Stay alert, keep calm, think clearly, and act decisively.

PRESCRIBED FIRE SITUATIONS THAT SHOUT "WATCH OUT"

- Burn unit not scouted and sized up
- Burn unit not seen in daylight
- Problem areas and potential spotfire areas not identified
- Unfamiliar with weather and local factors influencing fire behavior
- Uninformed on prescribed fire strategy, tactics, and hazards
- Instructions and/or assignments not clear
- Communication link not established with crew members/leaders
- Firebreak not constructed to mineral soil
- Lighting fire uphill or with the wind
- Attempting to burn strips that are too wide on backfire
- Unburned fuels in backfire area
- Cannot see back down the fireline; not in contact with anyone who can
- Burning within twelve hours of a frontal passage or predicted wind shift
- Weather getting hotter and drier
- Wind increasing and/or changing direction
- Having frequent spotfires cross the line
- Terrain and fuels too rough for pumper units to enter
- Stopping to eat lunch

Most of these situations are simple and require only common sense, but many times we forget this. Often we are in a hurry to finish a burn, or the weather conditions are not just like we planned, or firebreaks were not prepared properly. Whatever the issue, these situations, as well as numerous other scenarios, can cause accidents, injury, or even death to personnel. Always be as safe as possible, and do not be afraid to walk away from a burn unit if everything is not right. You can always come back and try it another day when everything is right.

FIRST-AID KITS

It is important that a first-aid kit be included on all vehicles used on the fireline. The American National Standards Institute has determined a minimum requirement for first-aid kits in the workplace (ANSI 1998). Prescribed burning activities fit under the type III classification for first-aid kits: used for outdoor applications, equipped with carry handle, have means to be mounted in a fixed position, and are corrosion resistant. These minimum standards apply for quantity, types, and minimum sizes of items needed in a type III kit:

- 1—Absorbent compress, 32 in^2 (207 cm^2); no side smaller than 4 in (10 cm)
- 16—Adhesive bandages, 1 in × 3 in (2.5 cm × 7.6 cm)
- 1—Adhesive tape, 5 yd (4.6 m)
- 10—Antiseptic, 0.5 g per application
- 6—Burn treatment, 0.5 g per application
- 2 pairs—Medical exam gloves
- 4—Sterile pads, 3 in × 3 in (7.6 cm × 7.6 cm)
- 1—Triangular bandage, 40 in × 40 in × 56 in (102 cm × 102 cm × 142 cm)

A first-aid kit should also include supplemental items specific to the area and type of work being done. It would be wise to include such items as insect sting treatment, eyewash, additional and larger burn-treatment dressings, specialized strip bandages, tweezers, and cold compresses. It is best to consult your local physician or medical supply dealer for more information on what to include in your first-aid kit.

Other Fire Safety Concerns

Be sure to have plenty of water or sports-type drinks available for the crew, especially if the weather is warm or the burn unit is large. Remember that if a fire escapes, personnel will need more drinking water, so be prepared. It is always easier to bring more water than is needed than to go back to town for more.

FOOD

Having food on hand may not seem to be a fire safety issue, but finding a way to feed the fire crew without compromising the safety of the burn is something that you will need to think about before you begin your burn. If you have a large burn unit that will take many hours to complete, or a unit that requires personnel to stay for several hours, you will have to plan on getting food to everyone. Napoleon said, "An army marches on its stomach," and a prescribed burn crew is no different. If the personnel do not have adequate energy, tasks become more difficult and take longer to perform. You must also plan a time for personnel to eat that will not interfere with burning operations. Often, once the backfire is in, several of the crew can take a short break to eat while other personnel patrol the fireline. This will give everyone time to take a well-deserved break and allow the fire to continue to back, creating an adequate blackened area before the headfire is set.

Under no circumstances should the crew ever leave the fireline to eat. A landowner once told me about a burn he was conducting. This landowner said that they started setting backfires on mowed firebreaks early in the morning; by noon his crew was hungry, so they stopped. They all went into the house and ate lunch. When he returned, he was very surprised to find that the fire had escaped. He had to call the fire department to help him get the fire under control. To this day that man still could not understand why that fire had escaped. Be sure that you always have people and equipment at the fire, because a burning fire needs to be attended.

EQUIPMENT

When working with equipment, be sure to follow all safety guidelines and procedures. Make sure all personnel know how to operate each piece of equipment they are assigned to use. This is essential for the person trying to operate the equipment, as well as for those people around the equipment.

When working around pumper units, be careful of hot mufflers that can burn. When refueling small engines, be sure to allow the muffler to cool down, use some type of funnel or pour spout to add fuel, and have a fire extinguisher available in case any gasoline fires start. Be mindful of electrical hose reels that have moving parts and of hoses with spray guns that can injure personnel who are not paying attention. The only time one of my crew members has broken a bone occurred when he was attempting to roll up the hose on a vehicle. His arm got caught in a hose coil, but the vehicle driver was not paying attention and drove off. The hose tightened around the crew member's arm and slipped off over his hand, but not before it fractured his wrist.

Keep hand tools sharp and covered when not in use. A sharp tool is easier and safer to use than a dull one. Be sure to always carry and handle tools in a safe and responsible manner. Hand tools should never be carried on your

Figure 9-2.
Properly using and holding hand tools are important for the safety of all personnel. Make sure personnel using them know how to use and handle all tools before conducting burns.

shoulder but near the head of the tool and down at your side (Figure 9-2). When traversing slopes, carry the tool at your side, on the downhill side of your body.

Wear proper eye and hearing protection when operating power equipment or when in close proximity to power equipment being operated. Be sure to wear eye protection when working in areas where hot embers are falling. When operating a chain saw during felling operations, be sure to have on the proper head, eye, ear, and leg protection. Also, have a spotter watching for falling embers and branches.

CLOTHING

Wear clothing that is safe, provides protection from heat and embers, and allows body heat to escape. If you have long hair, make sure it is either tucked inside your hat or helmet or tucked inside your shirt. People with facial hair should be extra careful when working with drip torches. On rough, uneven terrain, be sure to wear proper footwear; a minimum of 8-inch-high (20.3 cm), lace-up, lug-soled boots is recommended (NWCG 1986), but wear what is comfortable and safe. The main thing to remember is to carefully choose where you step, go slowly, and avoid hazardous situations whenever possible.

VEHICLES

When you are operating a vehicle, be on the alert for crew members working along the fireline. For safety, turn on the vehicle's headlights. Do not drive into thick smoke; you never know whom or what you will find on the other side. If necessary, place a person on foot to guide the driver. Be sure to watch out for gullies, ditches, wet areas, and steep slopes.

Be careful about stopping your vehicle in areas with extreme heat or where hot embers could float into the cab or back end where fuel is stored. Have a fire extinguisher available and within reach in each vehicle. Also, make sure that when you park a vehicle, you do not park in area of unburned fuel or on hot embers. Hot embers can cause tires to blow out, melt, and/or catch fire. I once had this happen with a four-wheeler at a burn. We parked two four-wheelers side-by-side in the black to go look over an area. When we returned, a tire on one of the four-wheelers had caught fire because it had been parked right on a smoldering cow chip. We lost a tire, along with the use of a four-wheeler for the rest of that burn, but fortunately, we put the fire out before the entire vehicle caught fire.

When backing up vehicles, use spotters to watch out for personnel and hazards behind the vehicle. Also, avoid tight turnarounds close to active burning or in areas of unburned fuel, as this can spell disaster if the vehicle becomes stuck or stalled. Make sure you will be able to turn around or make it through an area before trying it.

CHAPTER 10 Firebreaks

Even so the tongue is a little member, and boasteth great things. Behold, how great a matter a little fire kindleth!

—James 3:5

Firebreaks, also known as fireguards or firelines, are one of the most important elements of a properly conducted prescribed fire. Firebreaks serve several purposes, but the most crucial is to contain the fire within the boundary of the burn unit. Well-constructed firebreaks make burning safer and reduce the amount of work you have to do when conducting prescribed burns. Firebreaks should be constructed by removing any vegetation and exposing the bare ground or mineral soil so the fire cannot creep across the line and escape from the burn unit. Bare-ground firebreaks are by far the safest type.

One essential use for a firebreak is to delineate the boundary of the burn unit. This is vital so that members of the fire crew can see the boundaries and not light fires outside the burn unit. For example, we conducted a burn where the boundary around the unit had been plowed, except one area with quite a few trees. In that area, we blew a line down to bare ground with a leaf blower. We had an experienced person running the torch and stripping out the leaf litter, but the line was not well defined and the person was not informed where our firebreaks were. The person went about the assigned task and ended up lighting fires along what he thought was the firebreak. Fortunately, he was stopped about 50 yards (45.7 m) outside the burn unit, and we were able to get the fire contained quickly before any major problems developed.

Another important use for firebreaks is to allow vehicle and equipment access to the entire unit. Typically, vehicles constitute your main holding and suppression equipment. It is a great deal safer and easier on crew members if you can access the entire unit and not have to use hand tools to contain or suppress a spotfire. Firebreaks also allow vehicles to quickly and safely patrol the line while you watch for spotfires and problem areas. Having vehicle access is also extremely important for assisting with mop-up activities after the fire has died down.

While consulting on a prescribed fire at a ranch in the San Bois Mountains of southeastern Oklahoma, we encountered a situation that demonstrated the need for firebreaks around the entire unit. This ranch was just starting to use prescribed fire; the first unit they burned was about 800 acres (323.7 ha) and had very steep slopes. The manager claimed that there were about 300 yards (274.3 m) of fireline out of the 6 miles (9.7 km) total with a slope too steep for the dozer. The 300 yards actually turned out to be closer to 0.5 mile

(0.8 km). We had planned on making a firebreak by using rakes and leaf blowers. I told the manager that before the day was over, the ranch hands would figure out how to get a dozer down the mountain and make a firebreak because they would not like the hard work of hand lines. We started burning, and after three hours and two spotfires (with only leaf blowers and rakes to fight them), we had only 300 yards of fireline in. The manager and crew then understood how important it was to get a firebreak around the entire unit. When we finished eight hours later, the ranch hands knew that next time they could get a dozer down the slope. Good firebreaks assist with containing the fire, allowing access for equipment, and reducing the workload on the crew, which make setting fires a great deal safer and easier.

Firebreaks can also be used to reduce fuel along the edge of a prescribed fire area to make ignition easier and safer. For example, tallgrass prairies contain large quantities of fuel that are over 3 feet tall (0.91 m). This fine fuel can have flame lengths of over 6 feet (1.8 m) on a backfire, which can reach across a narrow firebreak. These flame lengths also create a tremendous amount of heat, which can put crew members at risk and damage equipment. In these instances you may need to mow or shred a desired width inside the burn unit and then cut your fireline to bare ground around the outside of the mowed area. This combination will help reduce flame length and fire intensity right next to the line, making ignition safer for your crew and equipment.

Another reason to mow inside the firebreak is that most prescribed fires are conducted just inside a fence line. When the flames are long and the heat gets extreme, the crew or vehicles have nowhere to go because they are pinned against the fence. This is something to think about when placing lines right next to fences. You are much better off mowing or shredding the heavy fuel loads along the fence line to keep personnel and equipment safe.

It may also be necessary to mow or shred inside the burn unit along the downwind sides of the firebreak to assist with the backfiring operation. Some plants, such as sand sagebrush (*Artemisia filifolia*), annual broomweed (*Xanthocephalum dracunculoides*), saw-palmetto (*Serenoa repens*), and sand shinnery oak (*Quercus harvardii*), are not completely burned during the backfiring operation, or there may not be enough fine fuels in their understory for a backfire to carry. This makes backfiring unsafe because unburned fuels can carry a headfire into the backfire area and across the firebreak. Changing the fuel architecture of these plants by mowing or shredding helps to ensure that they will be completely consumed next to the firebreak so the fire can safely back away from the fireline to create an adequate blackened area.

Firebreaks may not be difficult to build, but firebreak preparation is usually the most expensive part of a prescribed fire. So, once the breaks are established, they should be maintained to keep costs down. Using the firebreak as a road, or disking the breaks yearly, can maintain a firebreak that had to be created by a dozer. Lines that had to be dozed to remove trees or brush may need some type of yearly maintenance; you may want to disk, mow, or spray with the appropriate herbicide to keep the trees or brush from coming back between burns.

A major problem that occurs when conducting burns in areas with trees or brush is what to do with the slash piles when you prepare the firebreaks. Most people want to pile them right next to the line, but this is an accident waiting to happen. There are several things you can do to reduce the risk of an escaped fire from brush piles. First, push the brush outside the burn unit so that it is not a problem at all. The main drawback to this solution is that most of the time burns are conducted right on the fence line and you do not want to cut your fence numerous times to push brush through; or if the fence is the property boundary, you probably cannot put your piles onto the neighboring property. The next technique would be to scatter rather than pile the brush inside the burn unit. Individual trees produce fewer firebrands and are easier to mop up than large piles. Put the brush a safe distance inside the fireline. This safe distance will vary with fuel type and fuel load, but 100 to 500 feet (30 to 152 m) is usually recommended depending upon fuel type (Wright and Bailey 1982). Again, this may not be a viable option because the area may be timbered or already have high densities of brush. The farther the brush gets pushed back in the unit, the more piles that are made because these brush piles add to the brush that is already present. The situation becomes even more dangerous, and the cost of the firebreak also increases. If you are trying to keep the damage to your trees to a minimum, pushing the brush farther into the burn unit might harm even more trees.

To overcome this problem, you have two possible solutions. First, pile the brush right next to the fireline but doze a line around the pile to keep it from being in the burn unit. This way you can burn the unit without igniting the piles. Then you can come back later after the unit has burned and burn the piles so that the wind carries the firebrands into the blackened unit. The other solution is to grind up the trees. There are several right-of-way companies that use large grinders mounted on tractors or trac-hoes to clear trees for power lines (Figure 10-1). These machines turn large trees into small piles of chips, which are much safer to burn than large piles of whole trees.

Erosion Problems

When developing your firebreak, take care not to cause erosion problems. Many times on steep slopes you may not be able to take the firebreak down to mineral soil. Instead, you may have to use only a mowed line, which can still burn. A mowed line should be limited to as short a distance as possible and left for experienced crews.

If erosion is going to be a concern, you have several options to minimize the problem. One option is to take the time to make water bars or diversion dikes on the dozed or scraped lines to reduce channeling of the water. A second

Figure 10-1.
Using a shredder or mulcher along the edge of the burn unit to reduce the piles can make the burn safer.

option is to use J-checking on plowed or dozed lines, where the firebreak is diverted every so often in the shape of a J. This keeps the water from following a long, continuous path and diverts the water into or out of the burn unit onto the side slope, reducing the potential for erosion. A third option is to cut the firebreak right before the burn. This can be done the day before or the day of the burn. I do not recommend you do this while you burn, because if you have mechanical problems or get stuck with your tractor or dozer, you may not be able to finish the lines before the fire catches up to you. But if the lines are cut right before the burn and then rehabilitated following the burn, the ground should only be exposed for a couple of days, thus reducing most erosion potential. A fourth option is to plant some type of cover crop over your firebreak. Make sure the crop you plant is noninvasive, fire retardant, and green and actively growing when you plan on burning the unit. Many times people are also concerned about wind erosion from firebreaks on coarse-textured soils. The risk of not ever burning outweighs the concern that a small amount of soil might move; we need to remember that those sites burned historically under all types of conditions, and they are still there.

Width of Firebreaks

Many times recommendations state that the width of the firebreak should be ten times the height of the tallest flammable vegetation. For example, a person planning a burn calculates that the tallest trees in the burn unit are 10 feet (3 m) tall, then multiplies this height by 10, which equals 100 feet (31 m). The person then thinks that the bare-ground firebreak should be 100 feet (31 m) wide around the entire unit; this is not what the recommendation intended. Ten times the width of the vegetation means that an adequate bare-ground firebreak should be of adequate width, 6–10 feet (1.8–3 m); then you should backfire and blacken the remaining distance to reach the recommended 100-foot distance for safety.

The actual width of the firebreak will vary by vegetation type, region, burning conditions, crew experience, equipment, terrain, and agency policy. I have conducted burns on everything from cattle trails 12 inches (0.3 m) wide to bladed lines over 30 feet (9.1 m) wide. Numerous extension agency fact sheets, books, NRCS technical guides, and various other agency recommendations have been written on firebreak widths for specific vegetation types and regions throughout the United States. You should use these sources to gain the additional information for conducting prescribed fires in your area.

Types of Firebreaks

The following discussion explains the different types of firebreaks you can use.

NO LINES . . . NOT VERY SAFE

In some areas with steep slopes or large amounts of surface rock, you may not be able to use any type of prepared line or firebreak. If you are attempting to burn without a firebreak, you will want to have plenty of water, equipment, and personnel on hand because this technique is very unsafe. This type of burning should be left to experienced personnel.

Numerous successful burns are conducted on the Tallgrass Prairie Preserve (TGPP) for The Nature Conservancy in Pawhuska, Oklahoma, each year without a firebreak. The burns create a mosaic or patch-burn effect on the landscape. The fire crew try to use as many roads and natural breaks as possible when conducting these burns, but many times they have to go right across the open prairie. They do not want to blade or plow lines on the preserve; instead, they depend solely on wet lines to contain the fires.

The TGPP burns by spraying a wet line 4 to 10 feet (1.2 to 3.0 m) wide in front of a pumper truck and allowing one set of tires to run over the wet vegetation. This crushes the wet grass down to make the wet line more effective and gives the person lighting the fire a line from which to light. This wet line is followed by two to three pumper trucks carrying 800 to 1,200 gallons (3,028 to 4,543 L) of water; they have the ability to draft water from any water source in a very short time. This technique has come from trial and error, along with many years of burning experience. It works well because the TGPP has the personnel and equipment to conduct burns with this type of firebreak.

MOWED LINES/WET LINES

Mowed firebreaks are usually used in conjunction with wet lines. A wet line consists of water or foam sprayed on a portion of the mowed line, with the fire ignited just inside the wet line. The wet line serves as the firebreak, and the mowed line reduces the amount of fuel and fire intensity. Mowed and wet line firebreaks have been used successfully in different forms of topography and fuel types and are used extensively in erosion-prone areas (Dubé 1977; Martin et al. 1977; Wright and Bailey 1982). Even so, extreme care should be exercised when using these types of firebreaks because the fire can still burn through them.

To conduct a burn with a mowed line/wet line, mow down the perimeter of the burn unit as short as possible. On the day of the burn, spray water using a cone- or fan-type nozzle to a width of about 1 to 3 feet (0.3 to 0.9 m) in the mowed area. Just spray to wet the mowed fuel; do not waste water by completely soaking the area unless you have an endless supply of water. Next, light the fire directly behind the person laying the wet line and right next to the wet line in the dry area of the burn unit. Do not let the person with the torch get too far behind the person spraying the wet line, because the wet line may evaporate and become ineffective. Be sure to light the fire as close to the wet line as possible, and allow the fire to back away from the wet line an appropriate distance before stripping the fire out any farther. Remember that the only firebreak you have is a line of water. Be sure to have people come behind you to mop up along the wet line.

One way to burn from wet lines is to run two hoses from one pumper truck. The first hose sprays water right in front of the pumper truck. You want to try to spray the wet line right in front of one of the tires on the truck, which allows the truck to press the water into the vegetation and makes the wet line more effective. The ignition person follows behind the truck and lights the fires. The second hose is then stretched out 50 to 150 feet (15 to 46 m) behind the pumper truck (Figure 10-2). The distance depends upon the fuel load, flame height, and amount of heat being produced by the backfire. The person in charge of the second hose is there to suppress any spotfires and mop up along the edge. This person must be conservative with the amount of water used. Following the person with the second hose should be one or two people with hand tools and another vehicle patrol to look for spotfires or fire creeping across the line. Finally, use a leaf blower to completely mop up along the edge. The leaf blower can blow most smoldering mulch piles, cow chips, or small limbs back into the black of the burn unit.

Several problems can occur when using mowed lines. First, there is nothing to stop the fire from creeping across the mowed lines (Figure 10-3). What often happens is that you will be several hundred yards down the fireline when the fire burns through the wet line and across the mowed line; this then sets a fire outside the burn unit. If you go back to put out the escaped fire, you will often

Figure 10-2.
Putting down the wetline in the mowed area. Be sure not to get the fuel to be burned wet. This will slow ignition down and create safety hazards.

Figure 10-3.
The main problem with mowed lines/wet lines is the fire creeping back across the mown area and out of the burn unit.

run out of water because there is not as much water for suppression when the water is also being used for wet lines. Remember that it takes more water and more personnel to conduct a burn when using mowed and wet lines rather than bare-soil lines.

Second, mulch accumulates when you mow. Usually the wet line is placed right over some of the mulch piles; although the top of these piles does not burn, the fire can burn under the pile and escape from the unit. Most people on the crew will see a smoldering pile of mulch and just spray water on it before continuing down the fireline. When you come back later, the pile will be smoking or burning again, or the fire may have already escaped. If you have to use mowed lines, it is best to mow them right after the grass fuels go dormant in the fall if the burns are to be conducted in the late winter or spring. This allows for no regrowth of the mowed line and gives the mulch time to blow away or decompose before the spring burn. It also promotes the growth of cool-season plants that can help make a green barrier. The lines can be mowed during the summer months if livestock graze in the burn unit. The livestock should keep the mowed line grazed down and help reduce the amount of mulch that accumulates.

If mulch is going to be a problem, using a hay rake can make burning safer. Just set the rake to barely drag the ground, and it will clean the mowed line very well. An alternative is the use of a road sweeper with a nylon roller brush. The nylon brush allows you to go over rocks and stumps without damaging the equipment. If you rake the mulch, you need to decide on which side of the line to put the mulch. If you rake it into the unit, more fuel and intensity are added right on the line; on the other hand, if you place it outside the unit, more fuel is added to potentially aid an escape. The final decision is up to the fire boss and based on how much fuel is already in the burn unit and how much mulch is being raked up.

Another problem that needs to be addressed relates to spraying the wet line. Make sure the person running the spray gun does not spray the vegetation you are trying to light. This slows down ignition and creates skips due to unburned fuel, which can cause safety problems. Make sure the wet line is in the mowed area away from the fuel to be burned. If mulch piles are along the edge, make sure to move the wet line over so they do not get sprayed, as this could cause an escaped fire.

TRAILS—CATTLE/HIKING/HORSE

Trails make very good firebreaks, especially when used as a starting point for lighting the headfire. Trails can also be used for safely lighting backfires. It is best to use trails in combination with mowing, as this will reduce the flame length and fire intensity next to the line. If the fine fuel is reduced next to a 12-inch (30.5 cm) bare-ground trail, the fire should not be able to cross the line. But if the fine fuel is tall on both sides of the trail, be extremely careful because the fire may still be able to cross that line.

We conducted a burn using cattle trails on the backfire side for a rancher. We only had a short distance on either side of the area that had heavy fuels before we came to three well-worn cattle trails. We had access to a tractor and mower but opted not to use them because we were in a hurry. We started lighting the fire; the fine fuel load was light, and everything was going fine. The rest of our crew had gone the other way; they had a county road for a firebreak, as well as a wheat field on the other side of the road. We should have had the second pumper truck follow us on the side where there was the greatest potential for escape. When we came to an area with heavy fuel, the wind picked up and carried the fire right over the cattle trails and into heavy fuel across the fence on the neighbor's property. We could not get the fire stopped with only one pumper truck; we had to cut the fence and fight a spotfire that ended up burning about 40 acres (16 ha). It took both pumper trucks and about an hour out of our burning time, not to mention the toll this took on the fire crew, before we were able to put out the escaped fire. That burn was 1,000 acres (405 ha) and took us 3.5 hours to conduct. We could have finished a lot sooner and not worked as hard if we had taken 20 minutes to use the tractor and mower to shred the heavy fuel.

ROADS

Most roads make excellent firebreaks. First, they are already in place, so there is no cost in preparing them as firebreaks. Second, they are already bare ground and usually wide enough to prevent escaped fires. Several types of roads can be used as firebreaks. Two-track roads, or pasture roads, work very well, especially when setting the headfire from them, but they also work for setting the backfire. If a two-track road is used to set the backfire, make sure the road is well used and down to mineral soil in the tracks. Mowing is normally used in conjunction with this type of break. Often the center part of the road is covered in tall grass. If this is the case, you should mow the road before you begin burning. You may also want to mow one or two widths alongside the road on the side closest to the burn unit to reduce the height of the fuel.

Gravel roads and county-maintained roads make superb firebreaks (Figure 10-4) and do not involve any preparation cost or time. They are normally sufficient in width and allow for vehicle access. On the downside you may have to determine if the road is safe enough to work from. How much traffic is on the road? What time of day is it the busiest? Will it be safe for crews and equipment to be on the road? How much smoke can be put on the road before causing a hazard? These are questions you should answer before you ever start your fire. Most dirt or county-maintained gravel roads have very little traffic. If they do, it will usually be in the morning or evening, so plan your burn appropriately. You may need to plan for extra personnel to help with traffic control, or you could see if local law enforcement or county commissioner is willing to assist you.

Paved roads make good firebreaks as well but have two major problems. The first is obvious; paved roads normally have a lot more traffic on them, and the risk of causing an accident because of smoke or curious onlookers increases, not to mention the increased risk to the fire crew working right next to the road. The second problem is from firebrands sliding across the road. I have experi-

Figure 10-4.
Roads work well as firebreaks (photo by Stephen Winter).

enced numerous spotfires from grass leaves or hardwood tree leaves sliding across paved roads. The pavement acts like ice, and debris slides easily across the road. As with other types of roads, be sure to plan your burn according to traffic patterns. You may be able to request help with traffic flow and control by contacting the local supervisor of the state department of transportation or public safety office (Figure 10-5).

Many people are concerned about burning through fences when roads are used for firebreaks. Research has shown that repeated prescribed fires through barbed-wire fences did not affect tensile strength or galvanization of the barbed wire (Engle et al. 1998). Another study found that prescribed fires did not damage older corroded barbed wire (Engle and Weir 2000). Burning through fences can reduce maintenance by keeping brush suppressed in the fence line.

Although the barbed wire can easily withstand a fire, the wood corner posts may not fare as well. A lot of wood corner posts are old and have large cracks in them. This exposes more of the wood's surface, which allows the fire to burn into the post easily. If you believe that the wood corner posts are going to be a problem, treat them before you burn. You can weed-eat, use herbicide, or mow around the posts to reduce the chances of their catching fire. You may also

Figure 10-5.
Take extreme care when using paved roads for firebreaks. Traffic, smoke, and personnel on the road are major concerns that need to be addressed before burning.

want to spray the posts down with water right before you burn around them. Be sure to keep an eye on the posts so that if they do catch fire, they can be put out quickly.

Although prescribed fires do not typically affect the majority of wood line posts, they are a major concern because most prescribed fires are not conducted during extremely dry conditions, such as those under which a wildfire occurs. The burning of wood posts is related to 100- and 1,000-hour fuel moisture. A good example of where burning in area with wood fence posts has not caused problems is in Osage County, Oklahoma. Thousands of acres are burned there annually, and miles of wood fence posts have been burned through for several decades. One or two posts per mile might be lost in some years, but usually the posts suffer very little damage. The most important considerations are recent precipitation and the condition of the posts when you burn.

DOZED OR SCRAPED LINES

Some of the best firebreaks are ones that have been scraped by a dozer or maintainer. You want the equipment operator to scrape only the surface to remove the fine fuel. Done properly, these lines can be made relatively quickly and economically and will cause very little, if any, erosion. Dozed or scraped lines also provide a corridor for equipment and personnel to travel on safely and quickly. The main consideration for dozed or scraped lines is to find a reliable equipment operator who understands what you are trying to accomplish. Many times equipment operators are not accustomed to scraping only the surface and want to dig down several inches to make a large road. These roads more often than not resemble a moat around the burn unit. This type of operation can cause serious erosion problems and drastically raise the cost of the burn by increasing the time spent preparing the firebreak.

Double-dozed lines are another type of firebreak that uses dozing or scraping. In this method a firebreak is still placed around the entire burn unit, but you must plan which wind direction you are going to use when you burn the unit. Dozed or scraped lines are placed on the two downwind sides, inside the burn unit. The distances these lines are placed inside the perimeter are determined by fuel type and load. The recommended width is between 100 and 500 feet (30.5 and 152 m) depending upon vegetation type and fuel condition (Wright and Bailey 1982; McPherson et al. 1986; Bidwell, Weir, et al. 2003).

DISKED, TILLED, OR PLOWED

Disking makes very good firebreaks if done properly. The best method for preparing disked lines is to mow the firebreak area and then disk the fireline. You may need to disk over the lines twice in fine-textured soils. If this has to be done, it is best to go back over the lines in the opposite direction. The main concern with disking is to incorporate the herbaceous material and not have continuous fuel across the firebreak. Even light amounts of contiguous fuels can cause a fire to escape across a disked firebreak, so make sure there is no contiguous fuel left within the disked area. Disking works very well in coarse-textured soils; it leaves residue in the soil to reduce the impact of erosion and promotes the growth of many annual forbs utilized by wildlife (Figure 10-6).

Another way to make firebreaks is with some type of tiller. These firebreaks work very well and are used in the southeastern United States. The only drawback is that the soils need to have a coarse texture so the tiller can get into the ground, and the area needs to be free of large roots and rocks.

A single-bottom moldboard plow will work to make small bare-ground firebreaks. These lines are not very wide and should be used in conjunction with mowing in heavier fuels. These types of plows can cause erosion problems on slopes, so use with care.

FIRELINE PLOW

There are several commercial manufacturers of fireline plows. Most are made to pull behind dozers, while some will connect to the three-point hitch of a tractor. The plows will create bare-ground swaths from 24 to 48 inches (61 to 122 cm) wide depending upon the model. The depth of the plow line varies also with the size and depth setting of the plow. Smaller plows can cut furrows as shallow as 3 inches (7.6 cm), while larger plows create furrows 3 to 4 feet (91 to 122 cm) deep. Erosion can be a problem on slopes or shallow soils. Lines should be rehabilitated as soon as possible to reduce soil loss. Fireline plows have been used in the southeastern United States for decades and can be pulled

Figure 10-6.
The area in the center is the disked firebreak in sandy soil. The area on the right has been mowed down to make the backfire operation safer and to give the backfire a continuous fuel bed.

Figure 10-7.
A mowed fireline and plowed fireline after a prescribed burn. On the left side of the photo is a prime example of what happens when you do not burn.

over and through almost anything. The smaller plows work very well in conjunction with mowed lines for making firelines (Figure 10-7).

The smaller fireline plow works well on every type of fuel, but I have personally had some problems with the plow not wanting to dig into the ground with sod-forming grasses or dense unmown tall grasses. For the best results, mow before you use the fireline plow.

As with other types of firebreaks, on most burn units containing livestock, the livestock start using the plow lines as trails and can help keep the lines cleared to bare ground. This will reduce the amount of work and cost necessary to create firebreaks for future burns.

HAND LINES

In areas where dozing lines may not be practical or economical, or where it is too steep for equipment to travel, or where erosion may be a problem, hand lines may have to be used instead. These firebreaks should be kept to the shortest distance possible because they require a lot of labor and time to prepare. If leaf litter is the main fuel source, leaf blowers and rakes can be used to remove the litter and make very efficient bare-ground firebreaks. Again, these types of lines are labor intensive, so plan accordingly. The best technique for preparing lines in leaf litter is to have one or two leaf blowers blowing the leaf litter out of the way and have a couple of rakes following behind pulling limbs and logs out of the way. This type of bare-ground firebreak works well and can be made relatively quickly depending upon the terrain and fuel load. The person marking the fireline or running the lead leaf blower should follow the lighter fuels and use any type of natural firebreak, such as drainages and rocks, to make building these types of lines easier on the crew. If the area is covered in grass, hand line preparation becomes harder and is less effective. Care should be used when burning off these lines, and adequate water should be on hand for wet lines and suppression.

NATURAL BREAKS

Most natural barriers like bluffs, creeks, streams, rivers, and lakes make exceptional firebreaks. You should consider several factors when using these types of barriers. Make sure the barrier is wide enough so firebrands cannot cross it. One of the main things you should consider is what to do if the fire crosses the barrier. Can you get down, over, or across the barrier quickly enough to suppress the fire before it gets too large? You may have to place personnel on the other side of the barrier for added protection. This problem with a natural firebreak came up when planning a burn on Black Mesa, the highest point in Oklahoma. The Nature Conservancy wanted to burn the top of the mesa to control encroachment from one-seeded juniper (*Juniperus monosperma*). The edge of the mesa is very steep and has a large seam of exposed rock on the side, which seemed to make a good firebreak. Our major concern, though, was firebrands from the junipers being carried down the 400 feet (122 m) from the top of the mesa and starting fires on the sides. This would have allowed the fire to run up the sides and cause us to have two flanks of a spotfire to put out on very steep terrain, and getting vehicles around on the rocky flats at the bottom would not have been very easy. When using streams or creek beds, make sure the vegetation will not allow the fire to carry across and start a fire on the other side. Commonly, debris dams on creeks will allow fires to creep across if they are not removed. Dry creek beds or streambeds work fine, but again make sure the fuels are not sufficient to carry across a fire. Many times a creek or stream that may be dry for long periods will have large amounts of vegetation growing in the bed. Creeks make really good firebreaks, especially in the late spring, because during this time the bottoms along the creek become very green due to cool-season plant growth. This green belt makes an excellent firebreak.

When you use lakes or large rivers as firebreaks, the biggest difficulty is the smoke causing problems for people who are using the lake or live in the area. Plan your burns accordingly, and manage the smoke to reduce the impact. Having a large body of water next to the burn unit can make burning easier and safer and can allow for some creative burns.

Cultivated fields can also make prescribed burning very simple and safe. But you must be careful to make sure that no crops are planted in the field and that what is planted there will not burn. If crops are present, make sure you do not run a hot fire into the field. The heat from the fire can damage or kill the crops for a considerable distance. If crops are present, you may want to backfire away from the field a safe distance so you do not heat-damage the crop.

Snow can work as a firebreak also. The wildlife manager on a large ranch in Utah uses snowdrifts on the north side of the ridges for firebreaks in the spring. During the spring the snow will melt off the southerly aspects and the vegetation will dry out enough to burn. Workers ignite these areas and let the fire go to the top of the ridge where the snow is still several feet deep and serves as the perfect firebreak, as well as a good safety zone for personnel who have to get out of the way of the fires. You need to make sure the snow you are using is wide enough and is not just resting on the grass with a void between the ground and snow. Fires have been known to creep under light, narrow snowdrifts and burn outside the burn unit.

Another natural firebreak you can use is wet vegetation. This entails burning grass fuels that are scattered or surrounded by timbered areas when the leaf litter is wet. Most of the time grass fuels will be dry enough to burn one to three days after a rainfall event. You can burn the grass fuels into the wet leaf litter, allowing the fire to go out on its own. To ensure that there was enough rainfall to cause the leaf litter not to burn, you may want to run a test fire first. Once the grass fuels are burned and the leaf litter dries out, you can burn the leaf litter into the blackened prairie openings. The main thing to watch out for is any large amount of grass litter on the ground that did not burn, because the litter will dry out and carry a fire. Also, you may have a small surface fire burning over the already blackened prairie opening, which can cause an escape.

DAY-LIGHTING

Day-lighting is a technique used in combination with roads in forested areas. The biologist in charge of the Pushmataha Wildlife Management Area for the ODWC at Clayton, Oklahoma, uses this method and normally burns 20 to 30 miles (32 to 48 km) of blacklines each year with just himself and one other per-

Figure 10-8.
Day-lighting is a very effective method for burning in timbered regions. Day-lighting entails cutting all of the timber along the road to change the fuel composition. The herbaceous growth can then be burned when the leaf litter is damp, creating a blackened area for larger-scale burns to be carried out later.

son. The area consists of shortleaf pine (*Pinus echinata*) and oak (*Quercus* spp.); it has very little herbaceous understory when fire is removed from the system. If you burn and reduce the overstory, tallgrass prairie species abound.

Day-lighting entails cutting all the timber along the road to a width of 50 feet (15 m) or more (Figure 10-8). This changes the composition of the fuel along the road firebreak from a solid stand of timber and leaf litter to grass. To burn this area, wait for approximately two days after a rain in late winter. The grass is then dry enough to burn, but the pine needles and oak leaves are still damp and will not carry a fire. Most of the time the day-lighted area can all be burned as a headfire, making this technique extremely safe. By using this technique, you can blacken the lines around the entire burn unit and come back later when the conditions are favorable for burning through the forest litter. The first burn after the logging operation on the day-lighted area will be the most important for removing the logging slash. You may need to use an herbicide to control shrub and hardwood growth in the day-lighted area depending upon the fire-return interval.

VARIOUS OTHER TECHNIQUES

There are numerous other ways to prepare firebreaks. For example, cattle feeding grounds can be used for firebreaks. During the winter months when feeding supplement to cattle, try feeding the cattle along the boundary of the burn unit. You can move the feed area every few days until the cattle trample and graze down this area to bare ground, creating a firebreak that is 15 to 30 feet (5 to 15 m) wide.

Another way to utilize livestock for firebreak preparation is through the use

Figure 10-9.
Airboats are used to flatten the sawgrass around the burn unit for a firebreak; then the fire is ignited from the boat (photo courtesy of J. Schortermeyer).

of fences. For example, a rancher who planned a burn on 160 acres (65 ha) set up an electric fence about 50 feet (15 m) inside the boundary fence. During the winter months he made the cattle use the lane around the burn unit. Within a short period the cattle had trampled and grazed the lane vegetation down very short and made a nice firebreak, complete with a bare-ground cattle trail around the entire unit. Granted, this may not be practical for most people, but it worked well in this situation.

On another ranch, the ranch hands have used glyphosate herbicide to create firebreaks. This ranch is located in a region with a large amount of surface edge rock, and using a dozer can be very difficult, expensive, and destructive. The ranch aerially applied the herbicide along the boundary of the burn unit where there are no roads or too much rock to put a line in. One reason this method worked is that the headfire was set at the herbicided fireline; a bare-ground firebreak was created and allowed for minimum suppression problems because the fire was moving away from it.

Some type of drag can also be used to make firebreaks. I know of a situation in which an extremely large tire was dragged behind a vehicle to create a firebreak. The tire would wear down and break off the vegetation, thus creating an adequate firebreak; even though it had to be dragged around the unit several times, it achieved the desired effect.

Probably the most ingenious firebreak preparation method is one that is used in Florida. In the sawgrass (*Cladium jamaicensis*) sloughs in and around the Everglades, fire crews use airboats. They go into the swamp with the airboat and drive around the burn unit; this creates a firebreak by flattening the sawgrass and blowing water onto the vegetation adjoining the burn unit. They then light the prescribed fire from the boat; if the fire escapes, they drive the boat out around the spotfire, flattening the vegetation and wetting it down (Figure 10-9), which proves the adage that necessity is the mother of invention.

CHAPTER 11 Fire Equipment

Behold, all ye that kindle a fire, that compass yourselves about with sparks: walk in the light of your fire, and in the sparks that ye have kindled. This shall ye have of mine hand; ye shall lie down in sorrow.

—Isaiah 50:11

Equipment for conducting a prescribed fire can range from simple and low cost, to multifaceted and expensive. Many times it may be limited by budget, experience, or agency requirements. The main thing to remember is to use dependable, practical equipment that everyone on the fire crew can operate.

Protective Clothing

The first type of equipment that you need to think about for prescribed burning is clothing for personal safety. Many governmental agencies and personnel who manage private lands often disagree on this topic, but when working with private land managers, you cannot force or expect them to wear full wildland personal protective equipment (PPE). From working with private landholders for years and knowing other people who also do, I have learned that experienced private landowners will lose all interest in prescribed fire, along with all respect for you, if you show up to a burn in full PPE. In many locations you will be laughed off the fireline if you show up in yellow wildland fire gear, so you should plan your clothing accordingly. Every year, private land managers safely burn millions of acres with little or no PPE.

For safety, everyone should wear long-sleeved shirts made of 100% cotton or wool. Long sleeves are important for protecting your arms from burning embers, hot parts of a drip torch, and radiant heat, along with briars, thorns, and other things that can scratch you. Pants should also be full length and made of cotton or wool. These clothing articles should be free of rips, frays, tears, and holes (Figure 11-1). The fabric along the edge of these rips, frays, and tears can burn, although cotton or wool will not flame up but will burn like a wick. Also, any type of hole in your clothing will allow radiant heat or possibly flames to burn your skin. Do not wear synthetic fibers, especially nylon, because most synthetics will melt or are flammable. Be careful not to wear any clothing that you like while you are working on the fireline because most likely falling embers will burn holes in it.

If you choose to wear wildland fire–type clothing, use Nomex or Indura FR cotton. These clothes come in different material thicknesses, so pick the one that is right for your situation. When burning in warm-weather conditions, choose a shirt made of lighter-weight material so it will be cooler. I prefer Indura FR cotton over Nomex for two reasons. First, Indura FR cotton is a natural

Figure 11-1.
Clothing is a very important safety factor on any prescribed burn. Clothing should include long-sleeved shirts, long pants, and leather boots and gloves and should be made of 100% natural fibers or synthetic fire-resistant fabric.

cotton fiber that breathes and wicks away moisture. Nomex is a synthetic aramid fiber that does not breathe or wick any moisture away. Second, Indura FR costs much less than Nomex.

You might also choose to wear coveralls that fit over your regular clothes. You can purchase your coveralls new, or often you can find them used and at no cost. All the coveralls used by refinery workers or petroleum truck drivers are Nomex jumpsuits. These companies require their employees to change out their coveralls periodically. Many times these coveralls are in good shape and can be purchased for next to nothing, or you can ask to have them donated to your program. Another source for low-cost coveralls is military surplus through the Department of Defense's Defense Reutilization and Marketing Service (DRMS) or through military surplus stores. All pilot jumpsuits are made of Nomex and often are put into surplus with very little wear or damage. These jumpsuits can be obtained at no cost by federal and state agencies through DRMS or purchased by others at military surplus stores.

One point to remember about clothing is that two layers provide better protection from radiant heat and better moisture absorption than one layer does. On the other hand, two layers can cause an increase in heat disorders by trapping and holding more of your body's heat. In the late winter to early spring I normally wear a long-sleeved cotton T-shirt under my fire shirts to help keep me warm when I am not doing a lot of physical labor. This added layer will also help wick away some moisture when I exert myself. In the summer months I typically wear a short-sleeved T-shirt under my fire shirt to help wick away moisture. The short sleeves assist with the dissipation of body heat.

The next item is footwear, and wearing what is both comfortable and practical is important. Again, the main thing to remember is to burn in clothing that is comfortable and to know the dangers of wearing certain items. If you wear slip-on boots, make sure to keep your pants over the outside of them so firebrands or embers will not fall inside your shoes. It is best to wear all-leather footwear; do not wear rubber boots or pac-type boots. Federal agencies require their employees to wear all-leather, lace-up boots 8 to 10 inches (20.3 to 25.4 cm) tall with Vibram-type soles. This type of boot provides the best protection for your feet and ankles when working on uneven ground. Make sure your footwear is broken in before you begin burning; if not and you do a lot of walking during the burn, you may wear large blisters on your feet.

Another important item is a helmet to protect you from falling debris, embers, and equipment along the fireline. A helmet can also be used as a platform to hold your goggles, a neck and face shroud, and other items that you might need when burning. Helmets must be able to withstand heat, so choose one that is wildland approved by the National Fire Protection Association (NFPA); a lighter-weight and thinner helmet may end up dripping around your ears if placed in a situation with a lot of heat. If you do not wear a helmet, at least wear a cap or hat to keep embers from falling into your hair.

One very helpful piece of protective equipment is a shroud; shrouds protect the ears, neck, and face from heat. They are made of Nomex, can be lined or unlined, and attach by Velcro to the inside of a helmet. These shrouds can help keep the heat off personnel igniting fire in heavy fuel or in suppression of spotfires. One drawback to wearing full face protection is that you can be covered up so much that you feel you can withstand more heat from the fire than you really can, resulting in heat stress or burns.

Another helpful item is goggles to help keep smoke and debris out of your eyes. They are not 100% effective against smoke, but they are better than not wearing anything at all. Goggles are also very important to wear when a lot of ash or embers are falling. They are also necessary when operating or working around chainsaws or leaf blowers. There are numerous types of goggles out on the market; be sure to choose a pair that fits and is comfortable. Wearing eyeglasses can be a problem. Most goggles will not seal around the earpieces of the glasses or will put undue pressure on the earpieces, making them uncomfortable to wear. A couple of comfortable brands of goggles do have soft sides and seal well around glasses. A few manufacturers make prescription-lens goggles, although these are more expensive than regular goggles.

Next, you should have a pair of all-leather gloves; these gloves should not have wide cuffs and should be chrome tanned. Chrome tanning does not leave residues in the leather as oil tanning does. Most gloves state "oil tanned" on the tag, or you can feel the oils in the leather. Do not use these types of gloves because in case of burn-over or working in extreme heat, the oils in the gloves will heat up and cause the leather to shrink tightly on your hands. This can cause serious burns or even loss of fingers due to heat.

We encountered a spotfire on one burn, and a member of the fire crew had on a pair of gloves with leather palms, fingertips, and a leather strip across the back of the hand. The rest of the glove was canvas. While fighting the spotfire in heavy grass fuels, he was burned by radiant heat through his gloves. Every place on the back of his hand that was covered by canvas had second-degree burns, while the areas covered by leather were not injured at all. This is a good example of why you should choose an all-leather glove.

Numerous types of leather gloves are available; choose ones that fit and allow for freedom of movement. Other gloves have Nomex cuffs and linings. These insulated fire gloves offer great protection but are difficult to work in when performing tedious tasks such as adjusting the vent on a drip torch. On the other hand, they are very useful when mopping up after a burn. These thick, insulated gloves allow you to pick up hot debris and move it farther into the black without burning your hands.

Many people like to use a respirator or face cover to protect them from smoke. Smoke contains several hazardous components, but no breathing apparatuses are rated or certified for wildland or prescribed fires. Several face covers on the market will help reduce smoke irritation and the amount of particulate matter (PM) that is inhaled. Make sure these face covers are 100% cotton or Nomex; you do not want any soft plastics that can melt or metals that conduct heat around your face. Also, you want to keep bandannas dry so you do not chance getting a steam burn on your face.

Fire Weather Kit

One of the most important tools to have on a prescribed fire is a weather kit. This allows you to determine the current weather conditions on-site before, during, and after the burn. Knowing what is happening with the weather conditions will help you determine the most effective use of personnel and equipment. It will also help you monitor fire behavior as it relates to changing weather conditions.

Several digital handheld weather meters are now available. These meters can give you a variety of different weather conditions depending upon make and model. If you use a handheld meter, make sure to get one that at least gives you the temperature, relative humidity, and wind speed. Also, calibrate the meter to make sure it is accurate. I have worked with several digital meters that were not accurate, which is a concern because a false reading can be hazardous. When using these instruments, be sure to hold them in a way that produces accurate wind speed, temperature, and relative humidity readings.

The belt weather kit is one of the most accurate and time-tested weather kits available (Figure 11-2). This kit comes with a logbook, pencil, sling psychrometer, relative humidity slide rule, wind meter, water bottle, and compass. These weather kits can be carried on your person or in a vehicle and are not too difficult to use.

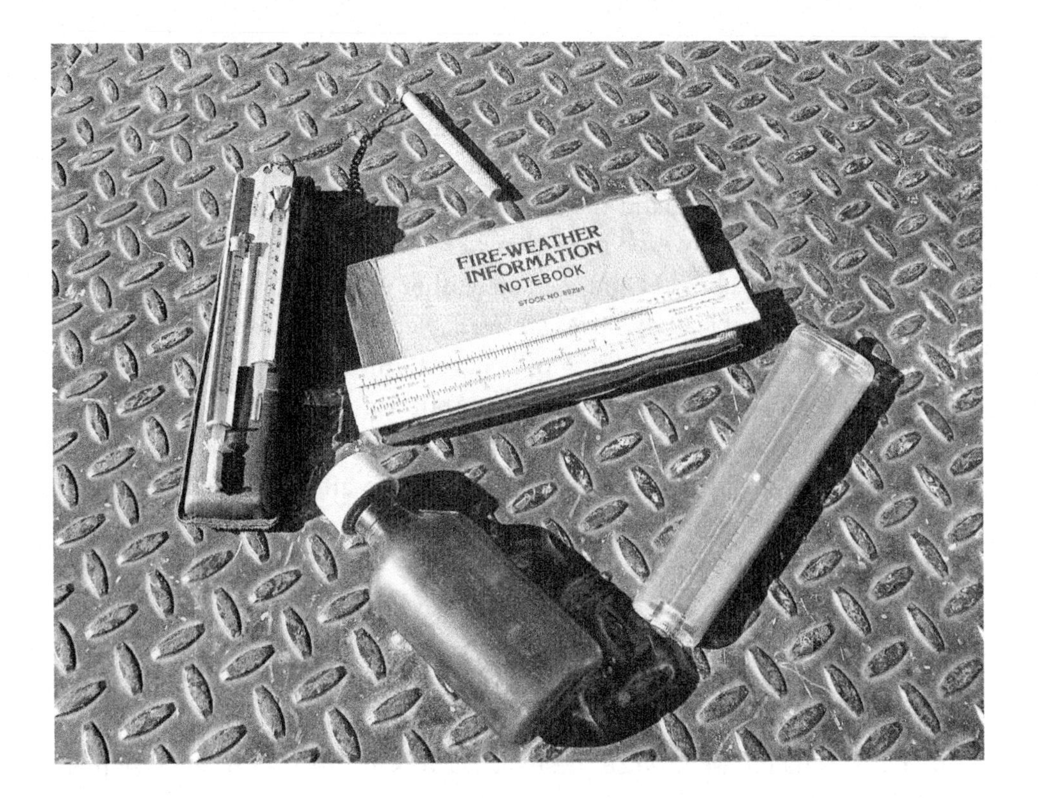

Figure 11-2.
A fire weather kit contains (clockwise from top) a logbook, slide rule, wind meter, water bottle, and sling psychrometer.

The logbook has spaces to write down time, elevation, and weather readings. It also has carbon paper to make duplicates of each page.

The sling psychrometer has two thermometers connected by a small chain to a small handle. The thermometer with the cotton wick is for wet-bulb temperature determination; the other is used to acquire dry-bulb temperature. To use the sling psychrometer, dip the cotton wick on the wet-bulb thermometer in the bottle of water included in the kit to thoroughly saturate it. Then grasp the handle, and twirl the psychrometer around for approximately thirty to forty-five seconds; do not twirl it too fast, or the wick will dry out and give you a false reading. Be careful not to hit the ends of the thermometers on anything and break them. When you have finished twirling, quickly read the wet-bulb thermometer first, then the dry-bulb, and record the readings.

Once you have the wet- and dry-bulb temperature readings, take the slide rule from the kit. There is a set of dry-bulb numbers at the top and bottom of the ruler and a double set of wet-bulb numbers in the middle. The top set of numbers is for dry-bulb temperatures between 20°F and 700°F (−7°C and 210°C); the lower set is for temperatures between 50°F and 1,000°F (10°F and 380°C). Determine which set of numbers to use, and line up the dry-bulb temperature reading with the wet-bulb temperature reading you have just observed. Then look to the right along that row of numbers until you see a set of small numbers with percent relative humidity above or below them. An arrow at the end of the wet-bulb numbers is positioned over the percent relative humidity. Record this number in the logbook under "RH." The ruler is set up in one-half degree increments, so make sure you line up the correct whole-degree numbers. If your reading does not seem correct, reread it or get someone else to check your numbers. If it still looks incorrect, retake your wet-bulb and dry-bulb temperatures.

Use the wind meter to measure the on-site wind speed. On the left side of the meter is a low-range scale from 2 to 10 mph (0.9 to 4.5 m/sec); on the right side is a high-range scale, from 10 to 66 mph (4.5 to 30 m/sec). To operate, face the prevailing wind and hold the wind meter at arm's length. Be sure to grasp the meter toward the bottom so you do not cover up the hole in the top of the meter that the wind passes over. Next, read the wind speed nearest to the white ball in the center of the meter. If the white ball is staying at the top of the meter, place your index finger over the red tube at the top of the meter and read the wind speed from the high range, right side. If the ball is not at the top, then use the low range, left side. The white ball will bounce with the wind and will probably not stay steady on one number. Just take an average over a several-second period. For example, if the ball is bouncing between 5 and 9 mph (2.2 and 4 m/sec), you will record the winds at 7 mph (3.1 m/sec). Do not let gusty winds increase your average; if the ball is steady between 5 and 7 mph (2.2 and 3.1 m/sec), then jumps to 10 mph (4.5 m/sec), record the winds as 6 mph (2.7 m/sec) with gusts to 10 mph (4.5 m/sec).

Hand Tools

Hand tools can be used to build firebreaks, mop up after the fire, suppress spotfires, and set fires. All hand tools require physical exertion to use and safe handling in the field. When carrying hand tools, you will want to find the balance point on the handle, usually just behind the head. Always carry hand tools on the downhill side of you when walking on slopes. Maintain safe spacing between other personnel when using hand tools. Do not stand with the tool resting across your shoulder; this can be dangerous to others if you happen to turn around quickly. If you have a sheath for the tool, you should use it to cover any sharp edges, not only for safety but also to maintain the tool's effectiveness.

FIRE RAKE/MCLEOD

The fire rake and McLeod, are each used by grasping the handle with both hands with the teeth down; the tool is then pulled toward you with desired downward pressure. Rakes are used to pull mulch piles from a mowed line before lighting, to scrape fuel away to get to bare soil for a firebreak, and to chop through small bushes and vines if needed. During mop-up procedures they can be used to pull smoldering debris, small limbs, and logs a safe distance into the burn unit. For direct-attack suppression purposes the rake and McLeod are best used in short grass and leaf litter where burning debris can be pulled into the black. For indirect-attack purposes, these hand tools can be used to make firebreaks down to mineral soil out a safe distance from the area of active burning.

Leaf rakes also work very well in certain prescribed fire situations. Be sure to choose a leaf rake with heavy metal tines. Do not use leaf rakes with plastic or bamboo tines because they will melt or catch fire, although the leaf rake can be used to clear a fire break, suppress spotfires, or mop up in leaf litter or light grass fuels. The flexibility of the tines makes leaf rakes very useful in rocky terrain because it is easier to clean debris from between rocks with a leaf rake than with a standard fixed-tooth fire rake.

SWATTER

The swatter, or flapper, is used by grasping it with both hands, placing it over the fire, and holding it there for a short period of time. It is best when used like a mop to smother flames. You can also extinguish flames in light fuels by swatting them. Be very careful when using this technique in heavier fuels because it will cause the fire to spread. Do not leave the swatter over the flames for a very long period because it can melt or catch fire. The swatter is best suited for direct-attack suppression in very short grass and leaf litter.

BROOM

Another hand tool that can be useful on the fireline is the fire broom. These brooms are lightweight and easy to use. The best use for a fire broom is to mop up along the edge of a burn unit. Use it just like an ordinary household broom, with a sweeping action to move smoldering debris inside the black. Be careful not to hold it over any hot areas for an extended period of time, or it may catch fire. Even though the bristles are treated with fire-resistant chemicals, they can still burn. Fire brooms can also be utilized to suppress low-intensity fire in light fuels. This can also be accomplished by sweeping the flames back into the burned area.

SHOVEL

The shovel is a time-tested firefighting tool. Shovels can be used to construct firelines to mineral soil and suppress spotfires. For suppression and fireline work, shovels work best in coarse-textured, loose soils and in areas where the vegetation is not extremely thick. I worked with a man from California who had been fighting wildfires there for over twenty years, and his favorite tool was a shovel. He came to Oklahoma for a fire school and asked us why we did not use any shovels. We tried to tell him that they just were not practical with our soils and vegetation. Nonetheless, he had to use his; in less than ten minutes he had put his shovel up, grumbling and complaining about tallgrass prairie and clay soils. Many times the best use for a shovel on a prescribed fire is to dig out a stuck vehicle, so it is best to carry at least one in each vehicle for such an emergency no matter where you are located.

BACKPACK PUMP

The backpack pump is available in several different configurations. One is a rubberized canvas bladder-type pack; another is a canvas pack with a replaceable bladder inside the canvas pack. Hard backpacks can be made of either plastic or metal. Whichever backpack style is used, they all hold 5 gallons (19 L) of water and weigh over 45 pounds (20 kg). The shoulder straps on most backpack pumps are not very comfortable. In fact, most of the straps are nylon and just over 1 inch (2.54 cm) wide. They dig into your shoulders and make the pump even more uncomfortable to wear. For many years we have replaced these shoulder straps with padded hiking backpack shoulder straps and added a hip belt to put most of the weight on the wearer's hips. Some pump manufacturers are now putting these on their new pumps. This configuration makes wearing a backpack pump more comfortable.

To operate the backpack pump, place the pump on your back and adjust shoulder straps for a comfortable fit. Grasp the wand in both hands, with your right hand holding the wand toward the rear and the left hand holding the bell-shaped nozzle end (if left-handed, hold opposite). Keeping your right hand and arm steady, pull out with your left hand on the nozzle end and then push it back in to discharge the water. This procedure is repeated as needed. The water should be aimed at the base of the flame to achieve maximum effectiveness.

Most backpack pumps have two spray nozzles. One has a single hole for a straight stream; this can be used for penetrating into soil or shooting water a farther distance. The other has two holes that produce a fan spray; this is used to rapidly cool down larger areas. The fan spray is the one that should be used most of the time because it is very effective for wetting down areas or fire suppression. If you need to change the nozzles, simply remove the one you are using, flip it over to the other nozzle, and screw it back down. Sometimes while you are using the backpack pump, the nozzle may become clogged from debris in the tank or water. If this happens, unscrew the nozzle and tap the bottom side against the wand to dislodge the debris. Before putting the nozzle back on, look through the holes to see if the debris was removed. If the debris was not removed, you may have to use a stem of grass or small twig to push the debris out. Care should be used not to break the twig off in the nozzle.

The backpack pump can be used to lay down wet line in areas where a pumper cannot reach or to follow up while burning from a wet line. Backpack pumps are also very effective for suppression of small spotfires. Backpack pumps should be placed along the fireline where they may be needed to suppress any creeping fires or spotfires. A person with a backpack pump can respond more quickly than a pumper can and can usually reach the spotfire when it is still small enough to be easily contained. Backpack pumps can also go into areas where vehicles cannot travel. If a spotfire occurs, the backpack pump can then be used to suppress fire as a main firefighting tool or as a follow-up tool to knock down flare-ups behind the pumper.

LEAF BLOWER

One of the most useful tools that I have ever used during prescribed fires is a leaf blower, because of its versatility. One of its main uses is in timbered areas to blow a clean line to bare mineral soil. One person running the leaf blower can make a trail, and people following up with rakes, McLeods, or shovels can clean any limbs or hard-to-move debris off the trail. Leaf blowers can also be used to blow out low-intensity backfires or flank-fires if the blowers are followed up with water and hand tools. Always be sure to blow any fire back on itself and into the black.

The leaf blower is best suited as a tool for mop up. A person with a leaf blower can be sent around the edge of the burn unit after the fire to blow any smoldering debris back into the black several feet. It works very well on mulch piles at the edge of the fireline by accelerating ignition and burning the material up, thus reducing the risk of escape. Remember to wear hearing and eye protection when working with or around a leaf blower.

Radio

A radio is a very important piece of equipment that should be used on every prescribed burn, and in a perfect world every person on the burn would have one. The lack of communication between crew members is the most dangerous safety aspect of any prescribed fire, and radios are important for allowing fast and easy communication between crew members.

Every crew member should be versed in the operation of each radio at the burn site and how to communicate using it. The best method for communicating is normal conversation or clear text. Be sure you remember to press the talk button before you begin speaking, and keep the talk button pressed down the whole time you are speaking. In emergency situations, do not shout; remain calm and talk slowly. Give specific directions or instructions so that everyone will know what the problem is and where it is located.

If there are not enough radios for the entire crew, be sure to spread the ones you have out along the fireline. The fire boss and crew leaders should each have one. Then spread the remaining radios along the fireline, making sure to have one at the back with personnel mopping up; this will ensure that if something rekindles or blows over the line, the crew can quickly get some assistance. Placing a radio in the middle of the ignition process can also be important while you are watching for spotfires. Another radio should be with or in close proximity to the person with the lead torch so this crew member can be notified

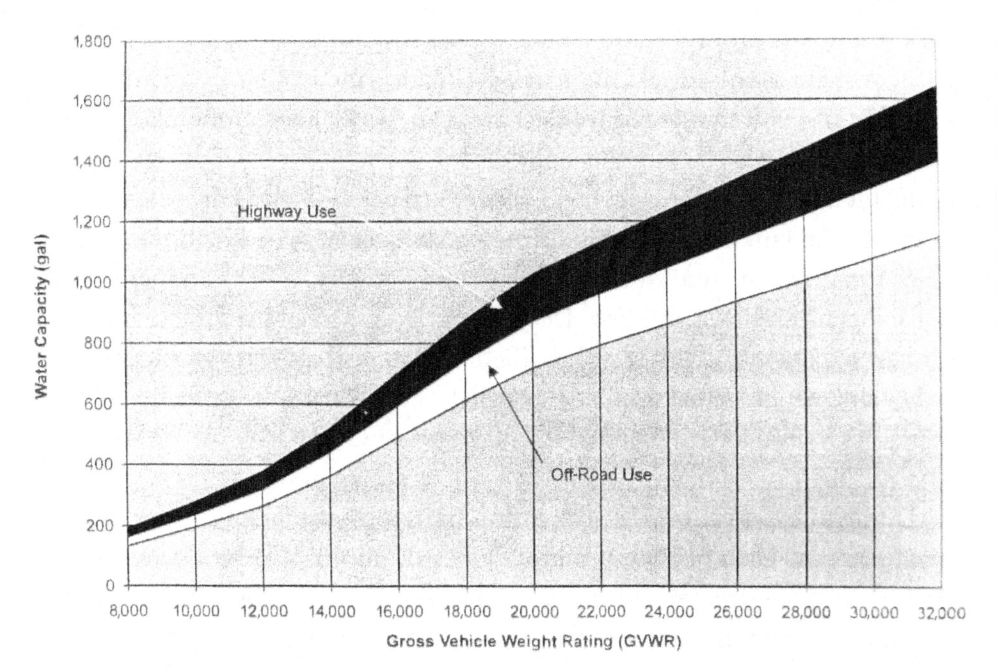

Figure 11-3.
The estimated water capacity for highway and off-road use of vehicles in relation to their gross vehicle weight rating. It is very important to put the correct-sized pump unit in the vehicle (from Roscommon Equipment Center 1990c).

if the fire is being lit too fast or if ignition should be stopped in the case of a spotfire.

Numerous types of radios are available. Make sure that the ones you choose are compatible with ones you already have or the ones being used by other people assisting you. Radios should have sufficient wattage to communicate throughout the burn unit. Remember that rougher topography means that the radios will be less effective. In cases of large burn units or burn units with numerous hills and valleys, it may be necessary to relay messages to other personnel. This is accomplished by relating your message to a person who can hear both parties that need to talk to each other. In some large burn units it may be necessary to position a person whose sole task is to relay messages between personnel. Make sure the batteries are fully charged or spares are available for those burns that take several hours.

Pumper Unit

Probably the most needed and costly piece of equipment for conducting prescribed burns is a pumper unit. These units can be shop-built for several hundred dollars or bought from dealers for several thousand. There are several items to consider when purchasing a pumper unit. The first is the size of the tank. Be sure to choose the tank size appropriate for the vehicle where it will be mounted. Do not overload the vehicle; know what the gross vehicle weight rating (GVWR) for each particular vehicle is. Overloading can be a serious safety hazard and can damage the vehicle. Most heavy-duty ¾ ton (0.68 MT) trucks are rated around 8,600 pounds (3,901 kg) GVWR. Figure 11-3 shows that this truck would rate around a 200-gallon (757 L) tank for off-road use. Remember that if you are going to use the truck off road, which you will for most prescribed burns, you should reduce the payload by 20% (Roscommon Equipment Center 1990c). Also, remember to consider the weight of the pump, engine, frame, fuel, and other tools that will be on the truck, as well as the weight of the water and the tank itself when figuring out tank sizes. The height of the tank is also important because it raises the center of gravity, which can increase the probability of a rollover accident. Depending on the size of the vehicle and tank, it is best to consider a tank that is baffled. Baffles will reduce water movement in the tank, which helps stabilize the vehicle and reduces the difficulty of stopping quickly with a tank that is not full. You should also have an ABC-type fire extinguisher located in each vehicle in case of emergency.

The next item to consider on a pump unit is what type of pump will best fit your needs. Each of the numerous types of pumps can produce different amounts of pressure per square inch (psi) (kg/cm^2) and flow rates in gallons per minute (GPM) (liters per minute, LPM). The two main types of pumps are high pressure–low volume and low pressure–high volume. A high pressure–low volume pump usually produces from 300 to 550 psi (21 to 39 kg/cm^2) of pressure and generates 8 to 16 GPM (30 to 61 LPM). These high-pressure units usually have a piston or diaphragm-type pump. A low pressure–high volume pump will produce from 50 to 250 psi (3.5 to 17.6 kg/cm^2) and flow from 20 to 150 GPM (76 to 568 LPM) and will usually use a centrifugal pump.

The type of pump you use will depend upon your requirements in the field. Most of the time water is the main limiting factor on prescribed burns. The most successful prescribed fires are ones where no water is used at all. But if water is needed, we want to use it to the best of our ability. A high pressure–low volume pump is recommended because these types of pumps conserve water and the high pressure also aids with suppression of the fire. If you have a high pressure–low volume pump that produces 10 GPM (38 LPM) and a tank that holds 200 gallons (757 L), you have the ability to spray water for twenty minutes. At the same time, a high volume–low pressure pump that produces 50 GPM (189 LPM) with the same size tank will give you only four minutes of water before you will have to refill.

A good example of why a high pressure–low volume pump is better can be seen from a wildfire training course involving several area fire departments. An area over 100 yards (91 m) long was to be ignited, and each fire department would work on suppression techniques. Some of the trucks had 200- to 400-gallon (757 to 1514 L) tanks and would be nearly empty by the time the fire was suppressed. To demonstrate the effectiveness of our high pressure–low volume pumps, our crew took a vehicle with a 50-gallon (189 L) tank and suppressed the fire using less than 10 gallons (38 L) of water. The point is that we still had water to use if another problem arose.

If a vehicle runs out of water, you must determine where the nearest water supply is and how long it will take to get there, fill up, and return. Can you continue to burn with this vehicle absent from the burn unit, or will you have to wait, losing valuable time and facing a potential change in weather conditions? These are also questions that you should ask when determining what type of pump unit is needed. Another question you should ask is, "What are the specific needs for a pumper unit?" If you are burning in an area with numerous buildings and structures that need to be protected, then you may consider having at least one pumper unit that could be used to extinguish a burning structure, or at least have the capability to wet down a large area quickly. In that case a low pressure–high volume pump would the most logical choice.

The next thing to consider is how you are going to get the water from the pumper unit to the fire. An important decision is the type, size, and amount of hose needed for a pumper unit. Hoses come in two types, hard line or soft line. Soft-line hoses are collapsible, light, and easy to maneuver, but the entire length has to be stretched out before pressuring up. These hoses cannot be rolled up while they have pressure on them. Soft-line hoses must also be dried out and cleaned before storage to prevent mildew damage.

Hard-line hoses have a hard outer shell and can handle higher internal pressures and external damage. These hoses can be unrolled and rolled up while pressured up. You can also pull out the amount of hose that is needed and use it. One drawback is that larger hose diameters become heavy, and it is sometimes difficult to pull long lengths of hose without help from another person. Even so, for most prescribed burning situations hard line is the best type of hose to use.

Hard-line hose with diameters of ½, ¾, and 1 inch (13, 19, and 25 mm) are probably the best sizes to use depending upon your pump's capability. If you are in areas where the terrain limits vehicular travel, you may want to have a hard-line hose that is up to 300 feet (91.4 m) long. This will allow you to reach areas where your vehicle cannot go, as well as allow you to have an adequate supply of water where it is needed. It is also useful to install a short hose or whipline. These whiplines are normally about 15 to 25 feet (4.5 to 7.6 m) long and used right around the vehicle, so you do not have to unwind or pull hose from the hose reel. The whipline also allows you to pull out the longer hose and work away from the truck and still work right next to the vehicle if the pump has the GPM (LPM) capacity to handle two nozzles working at once.

Another modification that can be added is a driver-operated hose and nozzle. A nozzle can be mounted on a boom in the front or rear of the vehicle with a control valve installed in the cab and operated by the driver (Figure 11-4). This eliminates the need for an extra person to run the hose outside the vehicle. These booms can be moved from side to side depending upon where they are needed. Most booms have to be moved manually, but they can be set up to

Figure 11-4.
This truck is equipped with a boom nozzle on the front for laying wet lines; the driver can operate it, thus releasing one person for another task.

operate remotely. The nozzles should be adjustable from a cone to a straight stream. There are now completely remote-controlled water monitors that are available and can be mounted on the outside of a vehicle.

Nozzles are also important and can range widely in GPM (LPM) output and cost. A nozzle with a variable stream adjustment works very well. Most adjustable nozzles will go from a straight stream to a cone or fog pattern. The cone or fog pattern works best for laying down wet lines, suppressing spotfires, and cooling areas down. The straight stream works well when trying to reach longer distances, knock down very hot areas, and penetrate into the soil to cool down areas during mop up. Be sure to choose a nozzle with the same GPM (LPM) as the pump. You are going to be extremely disappointed with the flow coming from the nozzle if the nozzle flows 30 to 50 GPM (114 to 189 LPM) on a pump that will only produce 10 to 20 GPM (38 to 76 LPM). It is very helpful to have a variable-flow nozzle to assist with conserving water. We have a high-volume pump on one unit with a nozzle that flows 10 to 30 GPM (38 to 114 LPM). We try to use it on the 10 GPM (38 LPM) setting as much as possible to conserve water, but when we need a large amount of water quickly, we can set it on 30 GPM (114 LPM).

FOAM

A foam unit is another helpful piece of equipment that can be added to a pumper unit. Foaming agents increase the efficiency of water many times over. Adding a surfactant, or class-A foam, to water produces numerous benefits: water penetrates and spreads better, water droplets are smaller, and flow is increased. The recommended ratio of class-A foam to water is between 0.1% and

1.0% (NWCG 1992). Foam can be very useful when laying wet lines that need to last for several minutes in dry climates; for protecting structures, trees, or other vegetation from heat and flame; and for smothering hot spots during mop-up.

Three basic foam mixes are wet, fluid, and dry (NWCG 1992). Wet foam is watery, with large to small bubbles. Fluid foam is similar to wet shaving cream and flows easily, with medium to small bubbles. Dry foam is mostly air and will cling to vertical surfaces but can be easily blown away by high winds. The consistency and amount of foam used will vary depending upon weather, need, and personal preference.

The foam solution can be mixed into the pumper unit in several ways. One way is batch or tank mixing. Batch mixing is just what it sounds like—mixing the right amount of foam concentrate into the tank. It is very inexpensive and simple to do, although it does have its problems. Some foam can cause problems with tanks, plumbing, and pumps; also, priming or cavitation problems can occur if the solution is too foamy when it goes through the pump. One of the main drawbacks is that the mix ratio cannot be changed very easily.

Proportioners can regulate the amount of foam concentrate added to the water. The foam is added to the line on the discharge side of the pump. These units can be adjusted manually or automatically depending upon the system. One of the main things to remember when using foam proportioners is to clean and flush the lines after each use so they do not become clogged.

After foam concentrate has been added by either method, foam is generated by using an aspirating nozzle. Aspirating nozzles use energy from the water pump to create the foam. The nozzle causes air to be pulled in, thus allowing the air and the foam solution to mix in the chamber and be discharged as foam. There are several different types of aspirating nozzles that will fit many different situations, so be sure to choose the one best suited for your needs.

Compressed-air-foam systems, or CAFS, are another way to produce foam. It is done by injecting compressed air into the foam solution. Using CAFS has many benefits: for example, the foam mix type can be changed easily, less foam concentrate is used, and the foam is consistently more uniform, which creates a better foam. On the downside, CAFS are a bit more complex, require a little more expertise to operate, and are more expensive than other foam systems.

Vehicle

The vehicle a pumper unit is placed in can vary as much as the pumper unit itself. The most effective and useful vehicles for pumper units are ¾- to 1-ton (680 to 970 kg) four-wheel-drive pickup trucks, although numerous other types of vehicles can and have been used for prescribed burning. Other types of vehicles, such as all-terrain vehicles (ATVs), utility vehicles, jeeps, tractors, and pumper units mounted on trailers, fitted with appropriate-sized pumper units can make burning safer and easier (Roscommon Equipment Center 1990a,

Figure 11-5. Pumper units and vehicles can vary in type and size depending upon needs and budget of your burn operation. The photos show four examples of pumper units that have been used on prescribed burns (clockwise from top left): converted military deuce and a half with 1,000-gallon (3,785 L) tank, tracked vehicle with 200-gallon (757 L) tank, Jeep with 55-gallon (208 L) tank, trailer pulled behind tractor with 300-gallon (1,136 L) tank and foam unit.

1990c; NWCG 1994b; Weir 2000) (Figure 11-5). Many types of military surplus vehicles have been outfitted for prescribed burning work, such as 2.5- to 5-ton (2.3 to 4.5 MT) trucks, gamma goats, and tracked vehicles (Roscommon Equipment Center 1976, 1988, 1990b, 1990c). The Roscommon Equipment Center, located in Michigan, is an outstanding source for information on equipping military and nonmilitary vehicles for fire use. This cooperative program between the National Association of Foresters and the Michigan Department of Natural Resources has also researched, tested, and developed guidelines and blueprints for vehicles, tanks, pumps, and many other items for fire use; more information can be found at www.roscommonequipmentcenter.com.

When thinking about using ATVs or utility vehicles, you must consider several items. These vehicles work very well for patrolling firelines during ignition and in the mop-up phase; they also allow personnel to patrol the fireline rapidly to watch for spotfires and problem areas (Figure 11-6). ATVs are faster and more maneuverable than utility vehicles, which can be beneficial, but sometimes faster is not better, especially if there is a lot of smoke and numerous person-

Figure 11-6.
Utility vehicles (left) and ATVs (right) are very useful on prescribed fires; they can be used for suppression, patrolling, ignition, and carrying equipment along the fireline.

nel are on the fireline. Make sure that ATV operators watch out for personnel on foot. The ATVs can also carry only a small amount of water for suppression of spotfires. On the other hand, the utility vehicles provide a fairly rapid mode of transportation and a platform that can carry two people with a pumper unit holding more than 50 gallons (189 L) of water. Because of their size, the utility vehicles can go into areas where normal vehicles cannot and can provide a sizable amount of water if needed.

Another way to acquire vehicles for conducting burns is to become a fire department. The Nature Conservancy's Tallgrass Prairie Preserve and several private ranches in Osage County, Oklahoma, have formed individual rural fire departments. This was done in conjunction with the Oklahoma Department of Agriculture, Food, and Forestry Rural Fire Protection program, which oversees all of the rural fire districts in Oklahoma. Becoming a fire department allowed these sites access to funds and surplus vehicles. These vehicles have been a great asset to their prescribed fire programs and have been valuable for suppressing wildfires.

Although it is important to use vehicles that are available or within your budget, do not place a vehicle on the fireline that will not handle the payload or terrain where you are burning. Furthermore, do not place a vehicle on the fireline that is too heavy or large for the burn unit. Also, make sure the vehicle is reliable and safe to operate. The only thing more discouraging than a vehicle that breaks down during a burn is one that malfunctions repeatedly.

Many people place pumper units on trailers, which allows the vehicle to be used for other duties when not being used for burning. One of the main drawbacks of using trailers is that they are hard to turn around very quickly. If you use a trailer-mounted pumper unit, make sure you have areas where the trailer can turn around rapidly if a problem develops behind it. Also, be sure the person driving can handle a trailer. Trailers cannot go many of the places a vehicle-mounted unit can go, which also limits their usefulness.

There are also other requirements for the vehicles used during prescribed burns. A vehicle should have enough space to get the fire crew to the burn unit. It is not safe to haul personnel to burns when they are riding in or hanging onto the back of a vehicle, especially if the burn unit is located several miles away. In addition, if the weather is cool, it is more comfortable riding inside a vehicle than outside, particularly after the burn is over. It may also be necessary to carry or haul equipment such as ATVs and utility vehicles, so extra trailers may be required. Do not forget to plan for these issues before they arise.

New types of equipment are being developed and marketed every year. Be sure to determine if these new types of equipment are reliable and will hold up to the rigors of burning. Be sure to keep the equipment you use simple so it can be operated by everyone working on the fire. Many times existing equipment can be modified to increase its effectiveness or to make it more user friendly. Be sure that any modifications you make to equipment do not void the manufacturer's warranty or make the equipment unsafe to operate. Sometimes you may have more equipment than operators, so make sure that each piece of equipment is needed before you take it to the burn site.

CHAPTER 12 Ignition Devices

So he drove out the man; and he placed at the east of the garden of Eden cherubim, and a flaming sword which turned every way, to keep the way of the tree of life.

—Genesis 3:24

To carry out a prescribed burn, you have to have an ignition source. Numerous items can be used to ignite prescribed fires. Some are very simple and inexpensive, while others are complex and costly. Choose the device that best suits your needs and your budget while allowing you to achieve the goals and objectives of your prescribed fire. I have been on more than one burn where everyone was briefed and in place, but when it came time to light the fire, the fire boss felt his pockets and had to ask if anyone had a match or a lighter, only to find out that no one on the entire crew had anything with which to light the fire. This shows why even the smallest detail is important when planning a burn.

Safety is the main concern when using any type of ignition device. According to an old adage, "If you play with fire, you are going to get burned"; if you are carrying something with which to start a fire, the chances are even better that you will burn yourself. Always be aware of your surroundings and the personnel around you when you are lighting fires so that you do not cause someone to receive a serious burn because of carelessness.

In the more than seven hundred burns I have been associated with, we have only had a few minor mishaps, but any one of them could have been severe. Once someone was not paying attention to what he was doing while carrying a drip torch down a steep canyon; as he worked his way down the side of the canyon, he allowed the torch end to drip fuel all over his lower leg. The fuel caught fire and burned the worker. Fortunately, the only injury he received was a 2-inch (5 cm) burn on his calf. On another burn, a person from a state agency was using an ATV and burning out areas that had been missed inside the burn unit. As he tried to cross a gully, the ATV flipped over backward, almost pinning him under it. The lit drip torch promptly ignited the plastic and foam seat, and the fire rapidly engulfed the entire ATV. The ATV and the radio strapped to it were both burned up. The driver received some minor burns on his arms while trying to put the fire out and a lot of ridicule from his fellow agency employees. This incident prompted the agency to require all their ATVs to have fire extinguishers mounted on them. Remember to be safe with, and have respect for, all ignition devices.

Matches/Lighter

The simplest ignition device, and one of cheapest, is a match or a lighter. Matches are lightweight, and a person can carry a large quantity of them. This ignition device will be used on most prescribed fires in one way or another. Even if you do not set the entire fire with matches, you will at least have to light something with them, such as a drip torch. Matches work well with heavy fuels and light wind conditions. In adequate fuels you can just strike the match, throw it down, and start the fire. In sparse or light fuels you may have to bend over a lot to keep the match lit. This makes for a long day and a sore back. Matches or lighters are not recommended for large burn units, although they work well for small areas, units that are totally protected from escape, and places where there is no need for much, if any type of, backfire. One of the main drawbacks to using matches is that you cannot make a continuous line of fire, which is very unsafe when setting the backfire. One ranch hand told me the best thing he could use was his horse and a box of matches, because the horse was not scared of him striking the matches and lighting fires as they rode around.

Drip Torch

The drip torch is probably the most widely used ignition device in the United States because it works well on all types of fuels and fuel loads. It also works well on different-sized burns. The drip torch has been used safely on prescribed burns for many years. With a drip torch you can put down as much fire as needed and even sling fire several feet into hard-to-reach areas (Figure 12-1).

To operate a drip torch, set it up from its storage position. First, unscrew and remove the lock ring on the top; then pull the spout from the torch body, holding it over the opening to allow excess fuel to drain back into the torch. Unscrew the flow plug at the bottom of the spout, and screw it into the flow plug holder so it will not bang around or come loose and fall into the torch body. Then turn the spout over, and set it on the torch with the loop facing opposite the handle. Replace the lock ring securely on the torch, but do not overtighten. Loosen the vent or breather valve, located on top of the torch above the handle, a half turn. Next, tip the torch down so that the fuel comes out. The fuel should be running out and hitting the wick, then dripping in a steady stream off the wick. To light the wick, you can use a lighter or matches; once the fire is started, if the wick goes out, you can just dip the wick into any flames to relight the torch.

To use the torch, tip it down to let fuel run out; when you want to stop lighting, just hold it upright so the flame is not near the fine fuel. To extinguish the flame at the wick, hold the torch upright and try to blow it out, taking care not to burn any facial hair; you can take your gloved hand and cover the wick to smother the flame; or you can use a combination of both blowing and smother-

Figure 12-1.
The drip torch is probably the most widely used ignition device for prescribed burning. It uses a mixture of gasoline and diesel for fuel.

ing. If you cannot get the flame extinguished, set the torch inside the burn unit in a blackened area and let the wick burn out. When you have to stop lighting for any period of time, it is best to extinguish the torch. This keeps the wick from burning up and helps reduce the amount of heat on the end of the torch. Sometimes the tip of the torch will get too hot and cause pressure to build up inside the torch body. The torch then operates erratically; either no fuel will come out, or the fuel will shoot out several feet. When this happens, it is best to stop and let the torch cool down for a few minutes before relighting it, or get another torch and keep going.

The drip torch is fueled with a mixture of diesel and gasoline. The ratio of the mixture will vary with ambient temperature and personal preference. A mixture of 50:50 diesel to gasoline works very well for temperatures ranging from 35°F to 75°F (1.7°C to 24°C). When temperatures are higher than 75°F, a mixture of 60:40 or 70:30 diesel to gasoline is recommended to reduce the volatility of the gasoline and to help the torch operate properly. When refueling a drip torch, keep it inside the burn unit. Move a safe distance away from any actively burning areas, and watch for firebrands that could blow into the refueling area.

When operating a drip torch, personal safety should be foremost in your mind. Personnel should have on gloves and a long-sleeved shirt that is either 100% cotton or a nonflammable material. People with long hair should tie it back and put it inside their shirts or under their cap or helmet so it does not get caught in the flames. A lit drip torch should never be allowed outside the burn unit. People operating the torch should be aware of the people around them at all times and not trap anyone with the fire or become trapped themselves.

To store the drip torch at the end of the burn, extinguish the flame and allow the wick to cool down; then reverse the setup steps. When you store the drip torch for extended periods, either remove the fuel or make sure the vent is closed and the lid and plug are fastened securely. If this is not done, the gasoline will volatize out of the mixture; the remaining straight diesel mixture will not ignite the next time the torch is used.

Fusee/Road Flare

Road flares, or fusees as they are more commonly called, are ignition devices that receive a lot of use within the federal wildfire programs. Because fusees are lightweight and portable, they are ideal for prescribed burns in remote areas or along firelines where vehicles cannot travel, where it is difficult to take in drip torches and fuel. Fusees produce a hot flame, work best on continuous fuels, and are more suited for small burn units. They will burn for about ten to twenty minutes depending upon their size.

To operate a fusee, you must wear gloves and adequate protective clothing. First, remove the cap, turn the cap half over, and strike the phosphorus end of the fusee with a downward motion, similar to striking a match. When you light the fusee, make sure to hold it away from your body and turn your head. Once it is lit, keep it from coming in contact with your body or clothing and watch out for the dripping phosphorus, which can reach temperatures up to 1,400°F (760°C) and will cause severe burns (USDI-BLM 1984). Once the fusee is lit, and depending upon fuel type and conditions, it can either be dragged through or dabbed into the fuel. Remember that a lit fusee must never be taken outside the boundaries of the burn unit. Fusees can also be thrown to ignite hard-to-reach areas. You should exercise care when throwing a fusee; do not hold it over your head or sling it in a careless manner because you or others can be hit with the dripping phosphorus. To extinguish the fusee, push the lighted end into bare mineral soil or gravel and hold it there until the fire is out. A fusee that has been extinguished can be reignited with another fusee.

As most fusees are only about 14 inches (36 cm) long, it is best to stick three or four together or use some type of extension handle to reduce fatigue to your back. This also provides additional safety by keeping the burning end farther from your body and out of your hand. When storing fusees, try to keep them in the original box and store them in a cool, dry place away from heat and flames. Fusees can deteriorate over time, so you will want to check them before each use.

Propane Torch

Propane torches are used by a lot of ranchers as an ignition device. The propane torch consists of a long metal wand with a hose of varying length attached to the propane cylinder. The torch works on sparse or continuous fuels and on light or heavy fuel loads. It produces a very hot flame, but there is no residual burning time as with drip torch fuel. One of the main reasons many ranchers use a propane torch is that they can ignite fires from a vehicle moving at fairly fast speeds without the torch being extinguished by the wind. One drawback is that the torch is usually connected to a large tank mounted in a truck or connected to a cylinder that is hard to carry by hand. Some people place the propane torch and cylinder in a basket and strap it to an ATV to allow for increased mobility along the fireline. Remember to exercise extreme caution when igniting fire from any type of vehicle. Another method for using and carrying the propane torch is to use the smaller propane cylinders and attach the torch to a metal backpack frame. The propane torch can then easily be carried almost anywhere an individual can walk (Figure 12-2).

Propane torches can also be used to reignite crown-scorched trees. A low-cost method has been developed for controlling eastern redcedar in tallgrass prairie areas following prescribed fires (Engle and Stritzke 1992a). This method requires a person using a propane torch to ignite live eastern redcedar trees that have been scorched several days earlier by a prescribed fire. This method has excellent results for controlling larger cedars not impacted by prescribed fire alone.

Pneumatic Torch

Pneumatic torches are similar to drip torches, except they are under pressure and can shoot flames 8 to 15 feet (2.4 to 4.6 m) depending upon the model and brand. Both backpack or handheld models work with either a mixture of diesel and gasoline or with straight kerosene. From personal experience, kerosene works best in this type of torch. Pneumatic torches work on all fuel loads and types, but they do not work very well in high wind conditions. The wind blows the flames away from and interferes with the fuel stream, causing difficulty with ignition. This type of torch works best when the fuel is fired with the wind. Pneumatic torches are a good tool to use when drip torches are not quite adequate or when you need to start a fire several feet away. Pneumatic torches also work well for reigniting cedars as described earlier (Engle and Stritzke 1992a).

To operate the pneumatic torch, observe all safety procedures for clothing. Make sure the torch is filled with the appropriate fuel; then pump the torch up or use an air tank if the torch is fitted with an air valve. Once the torch is pressurized, place the wick on the ground and spray fuel to saturate the wick. (If the wick does not get thoroughly soaked, the torch will not operate properly.) Then make sure you are inside the burn unit, light the wick, and test the torch to check that it is operating correctly.

To refuel, be sure the wick is extinguished; this can be accomplished by allowing the wick to burn itself out. Then make sure you are in a safe area inside the burn unit and away from all embers and actively burning areas.

Figure 12-2.
A propane torch is being used to ignite light fuels. Propane produces a hot flame but does not have a very long residual flame time. This propane torch is mounted on a metal backpack frame to increase its mobility.

Figure 12-3.
This ATV torch can ignite fire rapidly. Safety is of the utmost concern when strapping an ignition device to any vehicle. The terrain should be level and easily traversable. There should also be a fire extinguisher and an experienced driver on board at all times (photo by Stephen Winter).

ATV Torch

An ATV torch is an ignition device that mounts on the back of an ATV or four-wheeler and operates from the 12-volt system. The torch has a tank that varies in size, a 12-volt pump, and a pivoting arm that holds the torch end and can project a flame from 8 to 20 feet (2.4 to 6.1 m). These torches are commercially manufactured but can also be made in a shop; they use a normal mixture of drip torch fuel to operate (Figure 12-3).

There are some safety concerns with this type of device, mainly because it is strapped to the ATV. Extreme care should be exercised both while driving the vehicle and when igniting a fire. It is best to drive on a maintained firebreak that is fairly level. Extreme care and caution should be used when igniting across the burn unit to reduce the risk of turning the ATV over, causing injury, or losing equipment. You should have an ABC-type fire extinguisher on the ATV at all times. It is difficult to extinguish the flame at the end of this torch, so the ATV should never be driven outside the burn unit until the flame is totally put out.

The ATV torch works well in all types and quantities of fuels. It is best to secure the torch to the ATV with nylon ratcheting-type straps so that if the torch happens to catch fire, the operator can quickly cut the straps or allow the fire to burn through them. This will allow the operator to push the torch off the ATV safely and quickly.

Figure 12-4.
A marine flare gun and shells (top) and a hand-thrown flare with thirty-second fuse (bottom). These devices can be used to start fires in hard-to-reach or unsafe areas.

Hand Flares and Flare Guns

Hand flares and flare guns are used to ignite areas that are hard to reach or unsafe for personnel to ignite by hand (Figure 12-4). Flares work best on dry and moderate to heavy continuous fuels. A flare should not be used to ignite an entire burn unit, only parts of the area that cannot be reached by ground crews. They also can be used to ignite the center portion of a burn unit and help pull the fire from the edges to the middle.

Hand flares are 2 to 4 inches (50.8 to 152.4 mm) long and 1.5 to 3.0 inches (38.1 to 76.2 mm) in diameter. They normally have a twenty- to thirty-second fuse that must be lit by a match or lighter. To use, make sure all personnel are out of the area where you are going to throw the flare, light the flare, and throw into the area. Make sure the person throwing the flare can evacuate safely to the designated safe area. Most hand flares can throw sparks and flames up to 10 feet (3.1 m) and produce temperatures up to 4,000°F (2,204°C).

Flare guns can also be used for ignition. Most flare guns are made for marine or distress situations. Some flare guns are designed and manufactured for fire use and work well in moderate to heavy continuous fine fuels. Most flare guns can project a flare about 400 feet (121.9 m). The marine, or pistol-type flare gun, does not produce very intense heat or shower sparks over a large area, whereas flare guns made for fire use burn extremely hot and can ignite large areas.

To operate the flare gun, make sure all personnel are out of the area into which the flare is going to be shot. Keep the flare gun pointed in a safe direction, and do not load until you are ready to fire. Face the area, and shoot the flare into it.

Marine flare guns and flares are relatively inexpensive and readily available at marine supply stores or retailers that handle boating equipment. Flare guns and flares made solely for fire use are a better quality, but they are more expensive. They can be purchased from wildland fire equipment suppliers.

Branches and Hand Tools

Branches and hand tools work well when you have only a small area to burn. Typically, an area along the fireline will not completely burn out, and someone following behind the ignition crew or mopping up will find it. Instead of having a drip torch come back and ignite the area, you can collect a bunch of grass or leaf litter in the teeth of a rake, light it, and drag it along to ignite the fuels that have not burned. Be careful not to burn through the wooden handle on the rake. You can also use a branch from a coniferous tree to achieve the same results (Figure 12-5). These types of ignition devices work well on dry, moderate to heavy continuous fuels.

Terra Torch

A terra torch is a mobile flamethrower that can produce a lot of heat and a high volume of flame. Terra torches work well in all fuel types and fuel conditions but are most useful in areas that have light, noncontiguous fuels or dense stands of trees and shrubs that a surface fire will not burn through. Terra torches can shoot flames from 50 to 150 feet (15.2 to 45.7 m) depending upon model and manufacturer. They can be mounted in a vehicle or pulled behind on a trailer, but the perimeter of the burn unit must be accessible by the vehicle and trailer. The terra torch uses a mixture of diesel, gasoline, and a gelling agent. The mixture of fuel is usually 80:20, diesel to gasoline, and the amount of gelling agent depends upon both the manufacturer and environmental conditions. Cooler temperatures require less gelling agent but take longer for the mixture to gel. Warmer temperatures require more gelling agent but take less time for the mixture to gel. Learning the right mixture takes experimentation and knowledge of your equipment.

Safety is a priority when using the terra torch. At least two personnel should be at the torch, one mixing the fuel and the other operating the torch. All personnel involved with mixing the fuel and operating the torch must wear cotton or Nomex clothing. The personnel mixing the fuel should never use the torch, because any fuel spills or fuel vapor collecting on their clothing could ignite. You should have a minimum of two, 20-pound (9.1 kg) ABC- or BC dry-chemical-type fire extinguishers on hand for emergencies (Firecon n.d.). In some instances, you may need to rotate personnel due to a high amount of radiant heat. The operator should watch out for flaming fuel that may drip out

Figure 12-5.
Using a cedar branch or handtool to light an area that did not burn along the edge of a fire. This technique can be used on small areas when no other ignition device is available.

Figure 12-6.
A helitorch can produce hot fires and is useful when burning in limited fuels, difficult terrain, or large burn units.

of the nozzle and ignite any fuels underfoot. Read the operator's manual, and become familiar with the terra torch before using it. It is a very good tool for prescribed fire and natural resource management if used properly.

Helitorch

When terrain is complex or you have a large burn unit, the helitorch can be an extremely useful ignition device. A helitorch can deliver a high volume of fire in a short period of time over a large area. It can ignite standing fuels with very little understory fine fuel and fuels with high fuel moisture content (NWCG 1991). It also works well in remote areas or burn units with extreme topography (Figure 12-6).

A helitorch is a system that is hung externally from a helicopter. The torch consists of a tank, pump, and ignition device all attached to a frame and operated electronically from within the helicopter. There are many safety concerns when using a helitorch. First, you must have a helipad for refueling, which must be located within the burn unit. The helicopter cannot fly outside the burn unit boundaries because there can be malfunctions, such as the pump not shutting off or ignited fuel dripping from the torch. Either of these problems can start spotfires outside the burn unit. The helipad should also be located on the upwind side of the burn unit so the smoke will not hinder the pilot's visibility when landing and taking off. Being upwind will also reduce the risk of burning embers falling around the refueling operation. Personnel on the refueling crew need to be very careful and aware of their surroundings. They should be concerned not only with where the rotating blades are on the helicopter but also with the large quantities of fuel that are usually associated with helitorch operations. Make sure that all ground crew are aware of the conditions and watch out for themselves and each other.

Operation requires mixing the fuel, straight gasoline, with a gelling agent. The amount mixed in with the fuel is dependant upon the brand and environmental conditions. In colder weather, the gelling agent takes longer to gel, so you should not use as much as you would in warmer temperatures. If you mix in too much gelling agent, it will clog the pump and lines; if you do not use enough, the fire will be too hot and the fuel will burn up before it gets to the ground. It takes some practice, along with getting to know the equipment, before everything runs perfectly.

DAID/Ping-Pong Ball System/ Premo Mark III Aerial Ignition Device

The DAID, Delayed Aerial Ignition Device, or ping-pong ball system as it is more commonly called, is another form of ignition delivered by helicopter. The DAID can be used for large burn units, in remote areas, or in terrain difficult to traverse. It is a very effective ignition device in almost all types of fuels. It does not produce as much heat and flames as a helitorch, which is important if you are concerned about crown scorch in timbered situations. This is one reason the DAID is used so frequently in the forested regions of the southeastern United States. Using a DAID gives better control over the amount of fire placed on the ground, and the fire produced usually has a lower intensity (Figure 12-7).

The DAID works by injecting ethylene or propylene glycol into polystyrene spheres that are slightly smaller than ping-pong balls. Inside the sphere are potassium permanganate crystals. When the ethylene glycol and potassium permanganate are mixed, an exothermic reaction causes ignition (NWCG 1991). Ignition is not immediate; the delay depends upon how much ethylene glycol is injected. Typically 1 cc of ethylene glycol injected into the sphere will give about a thirty-second delayed ignition.

The DAID machine is mounted inside the helicopter and operates from the helicopter's electrical system. The operation requires someone to watch over the machine to make sure there are no problems and to keep the hopper full of spheres. You do not need as many personnel to operate the DAID machine as you need with a helitorch. A helipad is also needed on larger burn units for refueling the helicopter and for loading more spheres. This helipad does not need to be inside the burn unit because of how the DAID works. Safety is again a concern on the helipad; personnel need to watch out for the rotors when the helicopter is present and be careful when handling fuel during the helicopter refueling procedures.

Figure 12-7.
The DAID, or delayed aerial ignition device, is set up before being put into the helicopter. The top part is the hopper where the spheres are loaded; the middle part is the motor, injector, and dispenser. The piece in the foreground is the spout that attaches to the dispenser and hangs out the door of the helicopter.

Other Types

Other types of ignition devices are also used. Some people have used corn cobs soaked in diesel and gasoline stuck on a long handle. Others have tied an old tire soaked in diesel behind an ATV and dragged it around the burn unit as their ignition device. I know a rancher who has successfully designed and used his own version of a terra torch. Another rancher uses a propane torch attached to his truck with a lift arm controlled by a rope just outside the driver's door to lift the torch up out of the fuel. He has used this for many years and burns mostly by himself. The main thing to remember in developing and using an ignition device is to make sure the device is safe for the person operating it and the people working around it. Also, make sure it will do what you want it to do and that it is economical.

CHAPTER 13 Ignition Techniques

And call ye on the name of your gods, and I will call on the name of the Lord: and the God that answereth by fire, let him be God. And all the people answered and said, It is well spoken.
—I Kings 18:24

Prescribed fires can be ignited by many different techniques. You should ask yourself several questions before determining which technique is the most appropriate for a specific burn unit. First, will the technique allow for your specific goals or objectives to be achieved? Second, is it the safest method for burning that unit; and third, is it the safest technique for the crew and people around the burn unit? If the answer to any of these questions is "no," you should reevaluate the burn plan and use a different ignition technique that will allow you to safely achieve your goals.

Ignition Hazards

When igniting a prescribed fire, be aware of several hazards that usually occur along the edge of the burn unit while the fires are being lit and can cause the fire to escape the unit. First, watch for changes in the wind's speed or direction when you are burning; this is a common problem on any prescribed burn. Because wind speed and direction are usually not constant, the fire boss should be ready for a 45° shift in the winds on any burn; for example, if the wind is out of the southwest, you could very well expect winds to vary from the west to south. Most of all, know the weather conditions in the area where you are burning.

Entrapment of personnel is always a concern when igniting a fire (Figure 13-1). Make sure everyone completely understands the ignition plan. Have personnel watch out for the people directly in front of and behind them. Make sure all personnel are out of the burn unit unless they are lighting the fire. There may be a need to stop every so often and have everyone check in before proceeding with the burn. These precautions can help keep personnel from becoming trapped or injured within the burn unit.

Spotting, or spotfires, is another problem to watch out for. Spotfires can be caused by volatile fuels burning along the fire line (Bunting and Wright 1974; Weir 2007), heavy amounts of fine herbaceous fuels right on the line, fire or smoke whirls, hardwood leaves, or a variety of other problems (Wright and Bailey 1982; Weir 2007). Many of these spotfire causes can be managed by reducing the amount or type of fuel right next to the line or burning under the proper weather conditions (Weir 2007).

Slopes along the fireline can cause some potentially hazardous problems. Fire travels faster upslope, and the slope effect can override the wind direction (Pyne 1984). A fire burning up a 20° slope will burn four times faster than the

Figure 13-1.
Personnel can be trapped for a number of reasons, such as poor communication, lack of visual contact, or incorrect timing or staggering of ignition personnel.

same fire on level ground (Cheney and Sullivan 1997), so remember to be extra careful when burning on uphill slopes. The person igniting the fire and the suppression crew following the igniter may not be able to keep up with the fire. In this case the fire may race up the slope with an increased flame height and intensity, causing spotting, or the fire may even jump the firebreak. Probably the safest way to handle this situation is to stop igniting well before the foot of the slope so the fire will not back or carry up the slope. Then send a person to the top of the slope to ignite the fire back down the slope and tie in where ignition stopped. This fire should be less intense and easier to manage.

Heavy fuels can cause many problems along the fireline. An option for reducing your risk is to reduce the amount of fuel. This can be achieved by mechanical means such as mowing, shredding, chipping, mulching, or dozing. It can also be accomplished biologically through the use of herbicides or with grazing animals. Whichever method you use, make sure that it does not create more fuel or cause the fuels to increase in combustibility.

Another potential problem when backfiring a burn unit is incomplete burnout of blacklines caused by nonflammable, sparse, or irregular fuels, as well as breaks in those fuels such as livestock trails, pasture roads, rock outcroppings, or skips in the ignition pattern. To avoid incomplete burnouts, have the personnel following behind watch for areas that have not burned and burn out these areas as soon as it is safely possible to do so. Many times you will encounter fuels that will not carry a low-intensity backfire, but a more intense headfire would carry right across these fuels and could cause an escape. For example, common broomweed (*Gutierrezia dracunculoides*) can cause such a problem. In areas heavily infested with common broomweed the grass understory will burn, but the fire intensity may not be great enough to ignite the overstory of broomweed. When the next torch comes by lighting a strip headfire, the broomweed will ignite due to the increased intensity of the headfire, which causes the fire to carry right over the burned-out understory. Many other types of vegetation throughout the United States can cause incomplete burnouts. As mentioned previously, the best way to deal with this problem is to change the structure of the fuel through mechanical, biological, or chemical methods.

Irregularly shaped firelines can also cause major problems on a prescribed fire. Irregular shapes can cause a backfire to become a headfire that will burn intensely toward the unburned edge of the fireline. This problem typically occurs with personnel new to prescribed burning or on burns where the fire boss and crew are not completely familiar with the burn unit. To reduce this hazard, keep the firebreaks as straight as possible. When irregularly shaped lines are unavoidable, be sure to plan and think the burn out before you begin. Know where the irregular shapes are located, and stop burning a safe distance from them. Then redirect the suppression crew and ignition personnel to avoid any spotfires and to keep personnel safe.

Canyons can cause several problems when you are lighting fires. The first is a wind tunnel–like effect that usually causes increased wind velocity (Pyne 1984). As a result of the increase in wind speed, firebrands may leave the burn unit and start spotfires. The effect of the canyon can also cause the wind to change direction and be redirected up or down the canyon, resulting in erratic fire behavior along the fireline. The slopes of canyons can also have an effect on the safety of the personnel igniting the fires. Many times the canyon may be too steep to allow for ignition personnel to walk up and down. In these cases several options are available. One option is to stand at the top and sling fire from a drip torch down the canyon sides. Then once the canyon is burned out, the person can cross over. Another effective technique is to light the fire behind you while going down into the canyon. Once you have reached the bottom and the fire is out behind you, you can light the fuel at the bottom of the opposite side of the canyon and allow it to burn up to the top. Once it is burned out, climb up to the top in the black to be safe. If the canyon is too steep, use fusees or flares to safely deliver the fire down into the canyon.

The other hazard of canyons is getting suppression equipment from one side to the other. If you are not able to move equipment across the canyon, when you plan the burn, allow for extra equipment and personnel to be on the other side, or create an easier way for the equipment to get around the burn unit. If the equipment is on one side of the canyon and a spotfire ignites on

Figure 13-2.
For the safety of everyone on a prescribed fire, the entire crew should be aware of the ignition plan. The fire boss should make sure that each person understands his or her assigned task (photo courtesy Alaina Thomas).

the other, the equipment may have to travel several miles to get around to the fire. During this time the spotfire can increase in size and intensity, making suppression difficult. By already having the equipment on the other side of the canyon, you can minimize the risk of escaped fires.

Edges along vegetation changes can also cause problems along the fireline. For example, the edge where timber and prairie meet can cause winds to change direction and swirl (Schroeder and Buck 1970). Edges also cause differences in fire behavior that can catch fire crews off guard. In timber areas spotfires may not be a problem, but once you get into the prairie opening, the fine fuel changes and possibly becomes drier due to the amount of sunlight it receives. Remember to be cautious anytime you are dealing with vegetation changes within a burn unit.

Ignition Hazards to Personnel

Numerous hazards can directly affect the safety and well-being of personnel when igniting a prescribed fire. The main thing all crew members should remember is that watching out for their own safety and the safety of those around them will help reduce the possibility of any problems.

Poor communication is possibly the principal dilemma for personnel on a burn. Lack of communication can be caused by the fire boss or crew boss not explaining the ignition plan correctly, or the person in charge not being familiar with prescribed fire techniques (Figure 13-2). Whatever the reason, it can lead to a dangerous situation. The other communication problem is that not everyone running a drip torch has a radio and may not be aware of changes in the ignition plan or problems other personnel have encountered. If all personnel on a fire were able to have a radio, burning would be easier, but the cost of dependable radios usually does not allow for this. If everyone is not able to have a radio, then be sure to spread the radios out among the crew. Make sure that the lead, middle, and last drip torch operators have radios so they can communicate with the other torch operators in between them.

Lack of visual contact goes hand in hand with poor communications. Not being able to see the person in front of or behind you can also lead to precarious circumstances for personnel. If you are burning in areas where the terrain is extremely broken, or in thick brush or timber, you may want to have check points every 100 to 200 yards (91 to 183 m). This way you can make sure all personnel are where they are supposed to be.

Incorrect timing when sending torches to a burn area can also be hazardous to personnel. Sometimes one or two ignition personnel must wait for another person to ignite an area. If these ignition personnel do not wait, then you can have problems. This is the most dangerous situation because you will have people in the burn unit who can be surrounded by fire and be injured. Another timing problem may be that ignition personnel do not stay until they are supposed to proceed. If they proceed too early, they may set a fire without adequate black area to stop it and the fire could escape the burn unit. This is usually a problem when you have a complex ignition plan. Complex ignition plans are normally written by inexperienced fire bosses. Keep ignition plans simple, and know the experience and limitations of the crew and its leaders.

Using strip headfires or spot ignition techniques with incorrectly placed torches can endanger personnel. The personnel can be overrun or trapped by the person upwind of them. Preventing this problem goes back to good communication and having visual contact with the people directly in front of and behind you. Again, make sure all crew members understand the ignition plan and are watching out for the safety of themselves and others.

One sure way to endanger a person while lighting a fire is to have the wrong person running the torch. I have seen this happen numerous times. This situation has two extremes. First are people who are very excited about lighting fire, so they get a torch, take off at a brisk walk, and never look back. They will outrun the other torches and the suppression crew. The other extreme occurs when personnel are very timid, almost afraid of fire. These people will go too slowly, which may cause overrun problems for everyone. For example, during a training course we were conducting, a person was selected to operate a drip torch. The first thing he did was to walk outside the burn unit with the lit torch and start several small spotfires. Then, he was standing around with a lit torch, so I told him to put out his torch. He turned around quickly, flinging lit torch fuel on three people standing nearby. I promptly took the torch, and he never ran another torch around me. Remember to make sure the ignition personnel are competent and comfortable with burning before you turn them loose with a torch.

Another way to get someone hurt is being in a hurry to finish a fire. Many times we try to get a burn in before the weather conditions change or before normal quitting time. This hazard can be avoided by allowing plenty of time to execute the burn; remember to plan and allow time for the unexpected.

One situation you should watch out for is the burn crew becoming overextended or tired. Do not try to burn with a smaller crew than you actually need and end up having them spread out too thinly; this will quickly tire your crew. Also be aware that working on a hot summer day or having to suppress numerous or large spotfires can quickly tire out a crew. I recall one burn where we were burning blacklines late at night so the humidity would be adequate. We had just started when a large spotfire developed. We suppressed it and spent several hours mopping it up. We finished burning that night, but the entire crew was worn out. We made it back to town at 5:00 A.M. and returned to the burn unit at 7:00 A.M. to begin lighting the headfire. If another spotfire had occurred, we may not have been able to contain it because we were so tired from the previous night's work.

Good weather can also tire a crew because of the amount of burning that you might be tempted to try to get finished. Once we had really good weather conditions for an extended period of time, so we burned for nine days straight, burning nine different burn units. The entire crew was thankful it rained on day ten. Keep in mind that you need to pay attention to the energy level of your crew and avoid conducting burns when they are tired or when you do not have enough crew members for all of the assigned tasks. It is best to have a rule not to try to burn unless there is plenty of time for the burn, you have an appropriate number of crew members, and favorable weather is forecast for the entire burn.

Ignition Hazards within the Burn Unit

Many times hazards inside the burn unit's boundaries can injure crew members or cause liability problems. Make sure to check each burn unit for hazards prior to burning. If you checked the burn unit several months prior to the burn day, it is advisable to check it again closer to the actual burn to make sure that nothing has changed.

One of the main things you should be careful of is people within the burn unit, other than fire crew members, such as hunters, hikers, campers, joggers, horseback riders, or even homeless people. Other users can be especially problematic on public lands. Be sure to plan burns during times of minimal use, and post signs in the area warning of forthcoming burns. If possible, shut down the area until all the burns are conducted. Outside people within the burn unit can also be a problem on private lands. With many private lands being leased for recreational purposes, landowners often may not know when their lessees are on the property, so be sure to inform the lessees about the timing of any planned burns.

For example, I was involved with a burn on a state wildlife management area that was located next to a lake and had numerous houses around it. This area was supposed to be off limits, but many of the neighboring residents would jog, bicycle, or walk pets in the area. We planned a burn for this area, and before we conducted it, we had a local television news helicopter fly over the area for us to verify that no one was in the burn unit.

For safety and to avoid damage, you should locate all of the structures that are inside the burn unit, including all homes, barns, or other outbuildings. When examining these structures, determine what the risk of damaging them could be. It is important to determine if the structures are made of flammable or nonflammable materials and if they are old and will ignite easily. Old and unused buildings are harder to protect. You should also take inventory of the fuel load and fuel type surrounding any buildings to help determine what type of fuel reduction, firebreaks, and suppression equipment will be needed around the structure. If structures are located within your burn unit, you may need to have extra equipment and personnel on hand watching the buildings to make sure they do not catch fire.

A good illustration of this situation occurred while burning a unit that included an old abandoned house. The house was constructed of wood with a wooden shingle roof. The landowners wanted to save it, so we prepared a firebreak around the structure. The burn went well and did not get near the old house. The problem started when eastern redcedar trees ignited downwind of the structure and firebrands blew onto the roof. It ignited, and we did not have enough water or water pressure to put the fire out, so the house was lost.

Buildings are not the only structures you need to watch out for in a burn unit. Oil field equipment such as tank batteries and pipelines within a burn unit are a concern. I have never had problems with tank batteries that are being actively used. If they are being used or have been used in recent years, they usually have a well-worn road around them. Many times the road and turnaround area are graveled and have very little fuel in the area. At some tank locations you may have to mow and disk around the site to reduce the fuel load and keep intense heat away from the tanks. Remember that some of the tanks are fiberglass and will burn or melt from the heat.

If you contact the oil company, it will often clean up around the tank battery. The only problem I have encountered was around an old abandoned tank battery. This location had not been used in decades, and an old oil spill caught fire. The oil spill was not very large but had to be put out by shoveling soil over the burning oil. So check all tank battery sites for oil spills prior to ignition.

Gas and oil lines can also cause concern during a burn, although metal pipelines usually do not cause a problem. I have burned around numerous above-ground metal pipelines without any mishaps. Also, miles of metal pipeline in Osage County, Oklahoma, are burned through each year without any problems. Most state laws require pipelines to be buried. If there is a pipeline on a burn unit that you are concerned about, contact the company that owns the pipeline and request that the line be buried.

Plastic pipelines can be a problem when they are on the surface of the burn unit because prescribed fires can burn through them. The scope of the problem depends upon what is being carried in the pipeline. If a plastic pipeline is on the burn unit, ask the company to bury it or at least shut it down until after the burn is over. The company can then inspect it for damage to prevent leaks.

You should locate all dumps within the burn unit before you begin. The dumps have the potential to cause environmental pollution and to emit smoke that is toxic to personnel and the surrounding areas. Most of the time you do not know what has been put in the dump. It may also be against local, state, or federal laws to allow a dump to burn, and the person setting the fire that causes the dump to burn may be held liable. The result could be fines and/or jail time depending upon the law. A dump can also burn for several days after the rest of the fire has been put out, creating smoke problems downwind. All care should be taken to keep dumps from burning. Adequate firebreaks should be made around dumps, and suppression crews should be used to keep the dumps from burning.

Public utilities, such as electrical lines, can be another hazard in the burn unit. The main problem results from conduction of electrical current through smoke, caused by a dense smoke column rising through an electrical power line. The particulate matter and moisture in the smoke can cause the electricity to arc and ground out. Any personnel in this path can be electrocuted. If any overhead power lines go through the burn unit, make sure personnel know about them and are cautious while working around them.

The other problem with overhead power lines is the possibility of burning the poles. This can cause the lines to touch the ground or come close to the ground and create an electrocution hazard for crew members. It can also be an added expense if you have to pay for repairs to the power lines or poles. I have never burned a power pole, and I have conducted several hundred burns with power poles inside the unit. If the pole is old, or has a lot of damage (woodpecker, wind), and the conditions are extremely dry, a pole can catch fire and burn if not monitored. If there are power poles on your burn unit that are a concern, it may be best to mow or weed-eat around them to reduce fire intensity. Also, as soon as the fire has come through, have crews mop up and check all the poles. If a pole is on fire, be very careful about spraying water near the lines, as this can conduct electricity and cause injury to personnel.

You should also protect telephone junction boxes. They will not harm personnel, but fire will damage them. The heat from a fire can melt the wiring inside, causing surrounding neighbors to lose service and potentially lose confidence in your burning program. It may also cost your program economically by having to pay for repairs.

Ignition

When igniting a fire, you should always establish and start at an anchor point, an area where the fire cannot creep around behind or outside the burn unit. It is best to start in a corner if possible. If you have two crews, you can start in the middle along the appropriate downwind firebreak and send a crew in each direction.

When igniting the fire, have personnel light and walk into or across the wind. It helps keep the heat, flames, and smoke away from the people igniting the fire and keeps the fire from outrunning the ignition personnel and suppression crews, and possibly escaping from the burn unit.

Also, do not light too much fire at once, which can cause serious problems. Igniting an area too large for the suppression crew to monitor is the biggest problem. It may also create too much heat, which can put the crew and equipment in jeopardy. Furthermore, if enough heat is created, trees and shrubs may crown right on the fireline, causing spotfire problems. Also, make sure the ignition personnel do not get too far in front of the suppression crew. If the suppression crew has to monitor some areas and cannot follow up right away, have the ignition crew stop and wait until everyone is ready to proceed.

When you are widening out the backfire area and using a strip headfire technique, be sure not to make the strips too wide. The strips should only be about one to two times the width of the existing blackened area and firebreak (Figure 13-3). If the firebreak is about 10 feet (3.1 m) wide and there are 5 feet (1.5 m) of black from the previously lit backfire, then the next strip of fire can be lit out 15 to 30 feet (4.6 to 9.1 m) wider. The width depends upon several factors. One factor is the amount of fine fuel present next to the fireline. If the fuel load is heavy, then the strips need to be narrow and you may not even be able to go one times the width. If the fuel loads are light, you may be able to go wider than two times the width. Another factor is the weather conditions. Low humidity and high temperatures will cause extreme fire behavior; in this case, you should still use narrower strips, whereas high humidity and lower temperatures should allow for wider strips to be lit. The main thing is to watch the fire; do not let the headfire that was just set become so large that it will cross the existing blackened area and firebreak. If this is a concern, then narrow the strips.

For example, I have conducted fires where we used four drip torches to widen out the backfire and the fourth torch was only 8 to 10 feet (2.4 to 3.1 m) from the firebreak because the heavy fuel load was causing extreme flame lengths, even when only stripping out a few feet. I have also been in fuels where we were able to get over 50 feet (15 m) of blackline from only two strips. Make sure you

Figure 13-3.
A crew using a strip headfire to widen out the backfire. Make sure the people operating the torches do not overrun each other. The strips should not be wider than one to two times the width of the existing firebreak and blackened area.

pay attention to fuel loads and always observe the fire behavior behind you; you can then make the appropriate adjustments in the widths of the strips you are setting for maximum safety and effectiveness.

Another frequent problem when using a strip headfire is that the people igniting behind the lead torch do not allow the fire in front of them to burn down. The large flames created by the person in front of you should have reduced in intensity before you light your strip headfire. If you do not let the flames burn down, then the fire you are setting will combine with the one in front of you and create one large fire. This essentially is setting a fire without the protection of the strip headfire in front of you and can cause an escape. Also, if the strips are narrow, the heat from the fire of the person in front of you will be intense. There is no need for a person to walk next to a hot fire; wait until the fire has burned down or burned farther away before proceeding with your strip of fire.

Ignition Methods

A number of different ignition methods or patterns may be used when starting a prescribed burn.

BACKFIRE

Backfiring a unit consists of starting the fire on the downwind side at an anchor point, igniting the fire all along the fireline on the downwind side, and allowing the fire to back all the way through the burn unit (Figure 13-4). Caution should be exercised along the fireline until the fire has backed an appropriate distance, and suppression crews should be on lookout for firebrands. Once the ignition

Figure 13-4.
An example of a backfire. Ignite the downwind sides, and allow the fire to burn back across the unit.

personnel reach the corners of the unit, they should stop; if they continue igniting, they will be setting a headfire. As a safety precaution, the ignition personnel should not leave the fire but follow the fire as it backs up along the upwind firebreak. If the wind shifts, a person can light ahead of the fire to help keep the fire from becoming a headfire.

Backfiring has the slowest rate of spread and lowest fire intensity of all of the ignition techniques (Cheney and Sullivan 1997). Thus, it is a safe technique to use, especially for inexperienced personnel. It is a very good technique for reducing top-kill or scorch on trees (Biswell et al. 1973; Wade and Lunsford 1989). It also works well for the creation of a mosaic- or "patchy-"type burn for the formation of multiple wildlife habitats within an area (Smith 2000).

Backfires also create less smoke than other techniques because of the slow rates of spread and long residence time, which cause higher amounts of the fuel to be consumed in the flaming rather than the smoldering stage of combustion (NWCG 1985). Due to the low fire intensity there is usually less lift, so more smoke lingers at ground level. You will want to make sure that your backfires are conducted when weather conditions are favorable for good lift.

Watch out for several hazards when using a backfire. It takes a lot more time to burn a unit with a backfire than other techniques. During this extended burn time, many things can go wrong. The weather conditions can change and alter the fire behavior. The fire may not burn very well or become too intense. Wind shifts are another concern when using a backfire. If the goal of the burn is to reduce scorch on trees, make sure the weather pattern is stable and the forecast accurate. A wind shift can cause the backfire to become a headfire and result in mortality in many tree species or in the fire escaping the unit. You might want to consider burning in smaller blocks to reduce burning time. Smaller burn units could also help reduce the amount of damage to an area if a wind shift does occur.

HEADFIRE

A headfire moves with the direction of the wind and has been proven to be the most effective fire type for killing trees and shrubs (Fahnestock and Hare 1964; Gartner and Thompson 1972; Sackett 1975). Headfires are also more effective for burning through low quantities of fine fuel to reduce brush (Heirman and Wright 1973; Wink and Wright 1973) and will produce a more complete burn of a unit than other techniques will (Wright and Bailey 1982).

Complete burns like the ones headfires produce are often not as favorable for wildlife (Smith 2000), so be aware of the objectives of the burn. Headfires will also produce a higher smoke output (NWCG 1985) but create greater convection and lift. As a result, the smoke rises higher in the atmosphere where transport winds can carry it away from the unit. This greater convection will also cause the creation of fire and smoke whirls, which can cause spotfires (Wade and Lunsford 1989; Weir 2007).

Two techniques for using a headfire are most commonly used, ring firing and double-dozed lines.

Ring Firing

Called either ring firing (Packard and Mutel 1997) or the "Tallgrass Fire Plan" (Launchbaugh and Owensby 1978; Scrifes and Hamilton 1993), this technique is widely used throughout the Great Plains. The ring firing technique requires you to start a backfire at the farthest point downwind in the burn unit. At the same time you ignite the backfire along the downwind firebreaks in opposite directions. Personnel should stop at their respective corners before proceeding with ignition. This allows for adequate blackened area to be burned and makes sure that the other backfire is in place before proceeding. Once the backfires are properly in place, then the headfire can be lit along the upwind firebreaks. This allows the headfire to burn across the unit and stop when it meets the backfires (Figure 13-5). The width of the backfire area will depend upon fuel type, fuel load, weather conditions, and other fire-related variables (Launchbaugh and Owensby 1978; Wright and Bailey 1982; Scrifes and Hamilton 1993). If the backfire area needs to be widened faster, use the strip headfire technique.

It is best to start the backfire portion of this technique as early in the morning as possible. This allows the backfires to be put in under safer conditions: lower temperatures and higher humidity to reduce the possibility of escapes (Weir 2007). If nothing goes awry, the headfire will be set during the warmest and driest time of day. If the goals are for a complete burn, top-kill on resprout-

Figure 13-5. In ring firing a backfire is lit along the downwind firebreaks (top left), and then the flanks are ignited (top right). Allow the fire to back in the appropriate distance before igniting the headfire along the upwind firebreaks (bottom left). The fire should then pull together in the middle (bottom right) (photos courtesy Jay Kerby).

ing brush species, and total kill on nonsprouting species, use extreme prescription parameters for an intense fire.

One of the benefits of using the ring firing technique is that personnel do not need to go inside the burn unit, unless a strip headfire is used to widen out the backfires. It also requires less work and is less expensive than other techniques. The fire can, and should, be conducted in one day. I have personally used this technique on over seven hundred prescribed burns that have ranged in size from 1 acre (0.4 ha) to 3,800 acres (1,538 ha).

Depending upon weather conditions, fuel type, fuel load, and crew experience this technique works very well, although problems can occur. These problems can range from firebrands and actual spotfires to equipment malfunctions. Most of the time the problems occur during the ignition stage of the backfires. By the time the backfires are the appropriate width, the weather conditions are no longer favorable for the headfire. Then you may have to put the fire out and come back another day, or try to finish the burn the best you can in the time you have left.

For example, an 80-acre (32 ha) unit I burned with this technique took over eight hours; because of the dry conditions the eastern redcedar crowned and sparked numerous spotfires. The burn unit had only 1.5 miles (2.4 km) of fireline. The backfire portion of this burn took over seven hours. On the other hand, I burned a 3,800-acre (1,538 ha) unit with 14 miles (22.5 km) of fireline, and it was tied together in only six hours. On this burn we started early when the weather conditions were favorable and had no problems putting in the backfire; lighting the headfire took less than an hour. This example shows that the size of the burn unit may not be as important as the fuel type and loading, behavior of the fire, and weather conditions. But the size of the unit that can be burned with this technique is limited; it will vary with fuels, weather conditions, equipment, personnel, and burn unit.

Double-Dozed Lines

Henry Wright developed the use of double-dozed lines in West Texas (Wright 1974), and it is considered by some people to be among the safest methods for conducting prescribed burns (Scrifes and Hamilton 1993). It can be used on all types of fuels, fuel loads, and weather conditions (Wright and Bailey 1982). This technique works very well on large burn units, burn units with light fuel loads, and units with volatile fuels (Wright and Bailey 1982; Scrifes and Hamilton 1993).

This technique consists of making a bare-ground firebreak around the entire perimeter of the unit and then determining which wind direction is going to be used for the headfire. Once the wind direction is established, go to the downwind side(s) and create two more bare-ground firebreaks parallel to the perimeter firebreaks. How far inside the unit to create these lines depends upon fuel type, fuel load, weather conditions, and objectives of the burn. Consult local experts for recommended blackline distances before burning because these distances vary by region and vegetation type. To complete a burn with double-dozed lines, burn out in between these two lines first, creating blacklines on the downwind sides. It is safest to do this when the air temperature is low and relative humidity is high (Wright and Bailey 1982). When burning out between the two firebreaks, you can use the backfire, strip headfire, or flank-fire techniques. Remember to use the safest technique that allows for the most rapid completion of the task. Then in a few days or weeks, when weather conditions are favorable, ignite the headfire and allow it to run into the blacklines.

On large burn units where preburning of the blacklines cannot be accomplished in one day, it may be necessary to cut stopping points perpendicular to, and connecting, the inside and outside firebreak. This will give you a bare-ground area where you can stop the fire. These interior firebreaks can be put in at predetermined intervals, or you can use any natural breaks to stop the fire.

This technique has some limitations in areas with heavy fuels. In tallgrass prairie regions the mulch layer next to the ground will usually not be consumed when burning in blacklines under cool conditions. This can cause a problem when you set the headfire later because the mulch layer will have dried out and will burn again. All it takes is a firebrand to ignite the blackened area, and that area will reburn. The flame height is not very great, but the presence of volatile fuels such as eastern redcedar can cause a problem, especially if the cedars are browned out and dry from the blackline burnout. Even a low-intensity mulch fire can cause them to crown out and spot across the firebreaks. Therefore, you should be aware that this technique does have limitations.

FLANK-FIRING

A flank-fire burns parallel to the wind direction. The flank-firing ignition technique consists of lighting a fire into the wind and allowing it to spread at right angles to the wind (Wade and Lunsford 1989). This technique works well when you want to control the intensity of the burn or when a backfire would burn too slowly (Wright and Bailey 1982). It is also a very good technique to use when securing the flanks while using other ignition techniques (Wade and Lunsford 1989).

Flank-firing requires a firebreak to be placed around the entire unit. Blacken a sufficient area along the downwind side(s) of the burn unit; the blackened area will depend upon fuel type, fuel load, and weather conditions. Once the blackened area has been established and is the proper width, place ignition personnel along the upwind side of the backfire. The ignition personnel can be spaced somewhat evenly, or personnel in the middle can move closer to each other. Bunching the crew up in the middle of the burn unit will create a greater amount of heat and lift, pulling the outer fires into the middle. You should experiment to determine which technique is better for your situation.

Make sure that the people lighting along the flank-side firebreaks are right next to their respective firelines; if the wind shifts, the fire will not be able to make a large run at the firebreak and possibly escape. When the ignition personnel begin, make sure they stay abreast of each other. This configuration needs to be maintained the entire time, so make sure they see each other and at least remain in communication with the people on each side of them. If this is not possible, have check points at several intervals along the way and radios for each person so all personnel know where everyone else is. Flank-firing can be hazardous for personnel, so to minimize the risks, make everyone aware of the dangers. Anyone who falls behind will be in danger from the fires of the personnel on both sides coming together and trapping the individual (Figure 13-6). A wind shift can create a rapidly moving headfire that can injure or kill a person, so make sure that each person on the ignition crew is familiar with fire behavior and maintains visual contact with everyone else.

Flank-fires can cause convection that will increase the chance of fire or smoke

Figure 13-6. Flank-fire ignition involves lighting fires into the wind and allowing the fires to burn together. Make sure all ignition personnel stay abreast of each other (photos courtesy Jay Kerby).

whirls developing. As two flank-fires approach each other, they will intensify and start pulling together rapidly, at which point the convection and fire whirls typically occur. Convection can also cause intense fire behavior that may be too extreme for some situations. In timber areas it is best to use flank-fires with medium- to large-sized trees (Wade and Lunsford 1989).

STRIP HEADFIRE

A strip headfire is a series of headfires set progressively upwind of each other to rapidly widen out an area or to control the intensity of a prescribed fire (Wade and Lunsford 1989). Strip headfires are also used when a backfire would move too slowly and a headfire might be too intense and/or dangerous (Wright and Bailey 1982).

To conduct a fire using the strip headfire technique, prepare a firebreak around the entire burn unit and then ignite a backfire along the downwind side(s). The backfire should be allowed to back the appropriate distance for fuel

Figure 13-7. Strip headfires are used when backfires are too slow or a headfire will create fire behavior that is too intense for the objectives of the burn. Ignition personnel are sent across the burn unit, lighting fire at the proper distance to control fire intensity. Care should be exercised to make sure that ignition personnel do not get ahead of the person directly ahead of them (photos courtesy Jay Kerby).

load, fuel type, and weather conditions. Once there is enough blackened area, begin to ignite the strips (Figure 13-7).

If only one person will be lighting the fire, he or she should walk back and forth from each flank-side firebreak. Each time, personnel should make sure they ignite a strip that is the correct width for the prescribed fire objectives; the distance between each strip can be varied within a fire to compensate for variations in fuel loading or type, topography, or weather (Wade and Lunsford 1989). Ignition personnel should also be sure to light the flank sides each time, as shown in Figure 13-7. This will help ensure that the fire is contained within the burn unit if there is a slight wind shift. If you use more than one ignition person, make sure they are the appropriate distance apart based on the variations in fuel loading or type. For safety reasons, make sure ignition personnel do not get ahead of the person just ahead. It is also important that ignition personnel maintain communication with each other. Remember that these are headfires, and they can quickly overtake someone, causing injury or death. In areas where ignition personnel could be slowed down due to dense vegetation or rough terrain, it may be best to send only one person to light a fire at a time. Once that person has safely reached the far side of the firebreak, then the next person can proceed.

SPOT IGNITION

This ignition technique is known by several names: spot ignition (Wright and Bailey 1982), spot firing (NWCG 1991), area ignition (Fenner, Arnold, and Buck 1955), or point source fires (Wade and Lunsford 1989). This technique requires a backfire to be put in on the downwind side(s) of the burn unit. The distance the backfire should be allowed to back is dependent upon fuel type, fuel load, and weather conditions. Once the backfire is in place, have ignition personnel start lighting parallel to the backfire area. You will only light a small spot, then walk a certain distance and light another spot. This pattern is done all the way across the burn unit; then personnel can move upwind a predetermined distance and go back across the unit (Figure 13-8). Multiple ignition personnel can be used as in the strip headfire technique. For obvious safety reasons, make sure that personnel igniting the fire do not get ahead of the person downwind and just ahead. Be sure to ignite fires right next to the flank-side firebreaks in case the wind shifts, or have personnel set flank-fires along the sides to help contain fire. Timing and spacing of the individual ignition spots are key to a successful application of this technique (Wade and Lunsford 1989). Adjusting the spacing between ignition spots and the distance between ignition lines can help regulate the intensity of the fire. Also, adjust the ignition distances to regulate intensity for changes in topography, fuel type, and fuel load (Wade and Lunsford 1989). Because of the differences in fire intensities (headfire, flank-fire, backfire), this technique works very well when trying to create a mosaic-type burn for wildlife habitat.

OTHER TECHNIQUES

Center ignition is a technique typically used in areas with heavy logging slash and light winds (Beaufait 1966; Wade and Lunsford 1989). Center ignition involves igniting a fire in the center of the burn unit and then lighting the edges. For this technique to be successful, it is important to pile or push more fuel into the center of the unit so that an intense fire is created in the middle. With light or no wind, this intense-center fire should pull the fire from the edges to the middle of the unit.

Chevron firing is a method used to burn down slopes on ridge points (NWCG 1991, 1996). This technique uses a minimum of three ignition personnel starting at the same point at the top of a ridge. The lead person starts igniting from the top of the ridge at the end and works downslope. The other personnel then

Figure 13-8. Spot ignition is a series of ignition points along a line that can be adjusted by width and frequency of spots to regulate fire behavior and intensity. This technique is extremely useful in creating a mosaic burn pattern (photos courtesy Jay Kerby).

light downslope to the right and left of the first person, forming a chevron shape.

Another technique that works well when burning in backfires or blacklines is a modified flank-fire. I have used this technique quite a lot and call it "ninetying off." This technique works very well when personnel are limited and the wind is blowing parallel to or slightly across the firebreak from which you are attempting to backfire. To perform this technique, light along the edge of the firebreak for a short distance and then turn 90° into the burn unit and ignite a strip 5 to 50 yards (4.5 to 47 m) long (Figure 13-9). This ignited strip in the burn unit will burn like a small headfire and quickly pull the fire from the edges. You can rapidly burn out areas along the firebreak with only one person igniting fire. The distance a person lights before "ninetying off" and the distance a person "ninety's off" into the burn unit are dependent upon fuels and weather conditions. The main concern is not to create too large a fire at once. If the wind shifts slightly, the 90° strip will become a headfire. This technique also pulls

Figure 13-9.
This is a modified flank-fire technique called "ninetying off"; it only works when the winds are parallel to the firebreak. It allows for rapid widening of blacklines with only one person igniting.

the smoke and heat off the fireline, which makes it safer and more comfortable for suppression personnel.

Several other specialized ignition techniques work well. Many techniques use a hybrid of several different techniques to accomplish the burn. The main point to remember is to use the technique that is best suited to meet the goals and objectives of the burn and is safest for everyone concerned.

CHAPTER 14 Smoke Management

And he opened the bottomless pit; and there arose a smoke out of the pit, as the smoke of a great furnace; and the sun and the air were darkened by reason of the smoke of the pit.
—Revelation 9:2

Smoke is mainly a by-product of incomplete combustion caused by limited amounts of oxygen mixing within the combustion zone due to flammable gases being produced (Ryan and McMahon 1976). The amount of smoke produced is also distinctive for each of the four stages of combustion. (Mobley 1976; NWCG 1985; Hardy et al. 2001; Sandburg et al. 2002).

- *Pre-ignition stage:* Water vapor is expelled into the atmosphere.
- *Flaming stage:* The efficiency of combustion is relatively high; this stage puts out the least amount of emissions in relation to the amount of fuel consumed.
- *Smoldering stage:* The effectiveness of combustion is lower, resulting in greater particulate emissions. Studies show that the amount of particulate emissions produced per mass of fuel consumed during the smoldering stage is more than double that of the flaming stage.
- *Glowing stage:* This stage is characterized by minimal smoke because all volatile material in the fuel has been driven out and oxygen can now easily reach the fuel particle, making combustion more efficient.

Some concentrations of smoke are found in all phases of combustion, but they are greatest in the smoldering phase. Smoke is also more prevalent during the smoldering combustion of duff, decaying logs, and organic soils than in grass, shrub, and small-diameter woody fuels, which allows these fuel types to produce fewer emissions (Sandburg and Dost 1990). Also, the heat release of the smoldering stage is rarely sufficient to sustain convection, so the smoke stays near the ground in relatively high concentrations (NWCG 1985).

Smoke Emissions

Carbon dioxide (CO_2) is the largest single atmospheric emission from fire; although it is not considered an air pollutant, it is a greenhouse gas. Water vapor is the second-largest emission from fire; it is also not considered an air pollutant, but it does contribute to the total smoke load and causes reduction in visibility. Carbon dioxide and water vapor make up 90% of smoke emissions from fires (NWCG 1985; DeBano, Neary, and Ffolliot 1998).

Airborne particles, or particulate matter, are another major pollutant from prescribed burns; they can degrade air quality by reducing visibility, absorbing harmful gases, and aggravating respiratory problems in susceptible individuals (DeBano, Neary, and Ffolliot 1998; Hardy et al. 2001). A study from the Northern Territory of Australia found that during peak bushfire season the number of patients seeking treatment for asthma in Darwin increased as concentrations of PM increased (Johnston et al. 2002). The U.S. Environmental Protection Agency (EPA) (1998) maintains that if smoke intrusions lead to high PM concentrations, burn managers have two goals: to reduce the smoke's impact on public health and to take the proper steps to keep it from happening again. Also, PM collecting on the surface of objects can reduce the aesthetic appeal of and cause damage to these objects (Baedecker et al. 1991).

Carbon monoxide (CO), another by-product of fires, is usually found in higher concentrations when incomplete combustion of moist fuels occurs (DeBano, Neary, and Ffolliot 1998). The EPA lists carbon monoxide as a criteria air pollutant. Smoke from wildland fires is reportedly composed of hundreds of chemicals in gaseous, liquid, and solid forms, most of which are minor or of no concern (Reinhardt and Ottmar 2000). Following are some other main components of smoke emission from prescribed fire (Sharkey 1997c; DeBano, Neary, and Ffolliot 1998):

- Nitrogen (in the form of NO and NO_2)
- Sulfur (in the form of SO_2)
- Aldehydes: formaldehyde ($HCHO$) and acrolein (C_3H_4O)
- Organic acids
- Ozone (O_3)
- Semivolatile and volatile organic compounds like methane (CH_4)

Obstruction of visibility is one of the most objectionable features of the smoke from prescribed burns, along with how ominous the smoke column looks (Wright and Bailey 1982). Smoke-induced visibility problems can affect the safety of the personnel conducting the fire, public safety on roadways, and the recreational value of areas, as well as the public's reaction to prescribed burning programs in general. The environmental effects of smoke and other fire emissions range from localized visibility problems to possible changes in global climate (DeBano, Neary, and Ffolliot 1998). The EPA has also defined nuisance smoke as smoke in the ambient air that interferes with a right or privilege common to members of the public, including the use or enjoyment of public or private resources (Sandberg et al. 2002). Smoke management during prescribed burns is an important issue, so remember, "It's your fire, your smoke, and your problem." Be sure to think about whom and what your smoke is going to impact long before you strike the match.

Policy on Smoke

The Clean Air Act is the main law governing smoke and its emissions. First written in 1963, the Clean Air Act was passed by Congress in 1967. Since then it has been amended several times. In 1970, the amendments shifted air management and control from state priority emphasis areas to federal initiatives. As the states began to determine air pollution sources, smoke emissions from prescribed burning and wildfires came under scrutiny. In states where both population and smoke complaints were increasing, open burning policies began to be questioned. As a result, many states added open-burning amendments to their version of the Clean Air Act, with some states being stricter than others.

Many groups were also concerned with the air quality in rural and natural areas. In 1972, the Sierra Club filed suit against the EPA, requiring the EPA to deny approval of all state air quality plans that did not contain provisions for the prevention of significant deterioration in areas where the air quality was cleaner than the national standards (McMahon 1999). This led to an EPA policy on PM and the designation of National Parks and Wilderness Areas as Class I areas and the rest of the United States as a Class II area to help prevent any significant deterioration of air quality.

In the past two decades researchers have conducted numerous studies on smoke and its impact as it relates to prescribed burns and wildfires. These studies have led to federal and state policy statements on smoke and smoke management. In 1990, the Clean Air Act was amended again; prescribed burning became more widely recognized as a source of PM emissions and a major contributor to visibility problems in federal Class I areas (McMahon 1999). The EPA's interim air quality policy on wildland and prescribed fires states two goals: to allow fire to function, as nearly as possible, in its natural role in maintaining healthy wildland ecosystems; and to protect public health and welfare by mitigating the impacts of air pollutant emissions on air quality and visibility (EPA 1998). Numerous other people experienced in prescribed burning agree that in time, smoke and the lack of smoke management will severely limit or stop prescribed burning altogether in certain areas. Therefore, all people involved with prescribed burning must try to manage the smoke from their fires to the best of their ability. Fire managers should also become active in the policy and decision-making affairs of their local, state, regional, and national air quality regulatory agencies. This will ensure that adequate representation by the prescribed fire community will help protect our interests and that the personnel who write the policies actually implement and understand the practice of prescribed burning.

Smoke Management Programs

The Clean Air Act requires states and tribes to maintain national air quality standards to protect public health and welfare. The policy recommends that the states and tribes develop smoke management programs (SMP). These SMPS will help establish procedures and requirements for managing smoke from fires that are used to benefit our natural resources. The SMPS' purpose is to mitigate the nuisance and public safety hazards caused by smoke intrusions into populated areas, to prevent National Ambient Air Quality Standards (NAAQS) violations and the deterioration of air quality, and to address visibility impacts in federal Class I areas in accordance with regional haze rules (EPA 1998). The EPA also outlined indicators that determine if an area is in need of an SMP. These indicators include increase in citizen complaints about smoke; increase in monitored air quality values due to natural resource management fires; fires causing or significantly contributing to the monitored air quality that is greater than 85% of the daily or annual NAAQS of PM; or fires significantly contributing to visibility impairment in mandatory federal Class I areas. If a state determines a need for an SMP, it can implement any type of program that will keep from the area from having NAAQS violations and that will address visibility issues.

There is a two-tiered approach for implementing SMPS. Tier 1 is a voluntary program where agricultural burning for resource benefits rarely causes or contributes to significant air quality problems; tier 2 programs are voluntary programs implemented where agricultural burning would be expected to cause or significantly contribute to violations of the NAAQS or impair federal Class I areas (Agricultural Air Quality Task Force 1999). These SMPS may contain regulations that identify treatment techniques besides prescribed burning, specify basic weather parameters for burning and/or burn unit size, require voluntary notification or prior authorization before burning, and implement a compliance and enforcement system.

These plans are being implemented in many areas throughout the nation. For example, public land managers and regulatory agencies in Idaho and Montana formed the Montana/Idaho State Airshed Group to ensure that smoke dispersion and atmospheric conditions are closely monitored before prescribed fires are ignited on public lands. This will help ensure that air quality meets federal and state standards while prescribed fires are being conducted. Before prescribed fires are ignited, public land managers in Idaho and Montana submit their plans to the Montana/Idaho State Airshed Group meteorologist. The meteorologist reviews weather conditions and determines which prescribed fires can be ignited and which, if any, must be delayed to ensure that air quality meets federal and state standards. If air quality begins to approach unhealthy levels, public land managers may be asked to delay ignition (Montana/Idaho State Airshed Group 2005). Similar types of state and regional SMPS will soon be implemented across the nation, hopefully to the benefit rather than detriment of prescribed burning.

Smoke Dispersion Models

Smoke dispersion models can help burn managers estimate the possible smoke impacts across the landscape and the quantity of smoke being produced. Some states require a smoke dispersion output model to be included with each fire plan, so check with your state or local regulations. Local project managers can use smoke dispersion models to supplement the information required to obtain burn permits or to assist with visualizing the smoke impacts of different burn scenarios; regional fire managers can use them for environmental impact

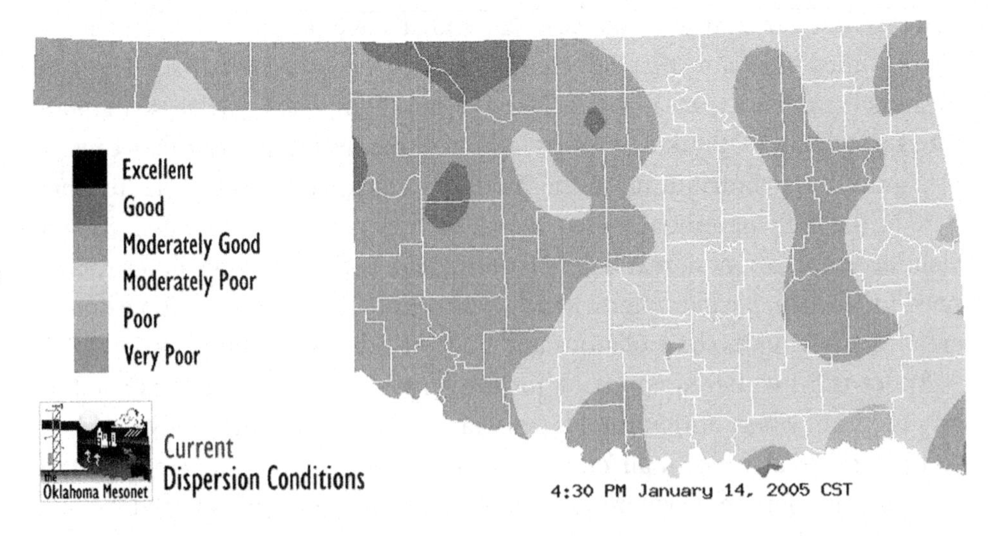

Figure 14-1.
This Oklahoma Mesonet smoke dispersion model shows the current dispersion conditions as well as forecasted conditions (http://agweather.mesonet.ou.edu/data/public/mesonet/models/realtime/dispersion/current.disp.gif).

assessment on a regional level or evaluate the trade-offs between prescribed burning and wildfires (Breyfogle and Ferguson 1996).

Several smoke dispersion models have been adapted or designed specifically for prescribed and wildland fires and are available on the Internet. The Oklahoma Mesonet system model can be found on its Web site at http://agweather.mesonet.ou.edu/rangeland/default.htm. The Oklahoma Mesonet dispersion model is a management tool developed to aid in decision making with respect to the near-surface release of gases and small particulates (airborne particles with diameters less than 20 ų). This dispersion model provides a current (updated every fifteen minutes) dispersion map for the entire state of Oklahoma, as well as forecast maps linked with the latest sixty-hour forecasts to produce maps of future dispersion conditions at three-hour intervals (Bidwell, Weir, et al. 2003) (Figure 14-1).

The Florida Division of Forestry maintains a smoke dispersion model on Florida's Fire Management Information System Web site at http://tlhforweb2.doacs.state.fl.us/SST/#SST. This site is a free-access Web-based smoke screening model that uses computer technology and forecasted weather data to view the potential impacts from a smoke plume. It is user friendly and can be used by prescribed burn managers within the state of Florida to plot smoke trajectory and identify local smoke-sensitive areas (SSAS).

Another Web-based smoke model is BLUESKY, located at http://www.fs.fed.us/bluesky; this model is being developed by the U.S. Forest Service AirFIRE Team under the National Fire Plan. BLUESKY is a real-time model designed to predict smoke impacts from wildfires, prescribed fires, and agricultural burning. It combines the most up-to-date fuels, combustion, emissions, meteorology, and dispersion models to create the best possible predictions of smoke impacts. The following PC-based dispersion models are currently used by Federal land managers: SASEM (Sestak and Riebau 1988); VALBOX (Sestak, Marlatt, and Riebau 1989); VSMOKE and VSMOKE-GIS (Lavdas 1996); NFSpuff (Harrison

1995); TSARS PLUS (Hummel and Rafsnider 1995); and CALPUFF (Scire et al. 1995).

Guidelines for Prescribed Fire Smoke Management

The main goals of smoke management are to reduce emissions from a fire, improve the dispersion of smoke columns, and make sure smoke plumes do not affect smoke-sensitive areas. You can use a five-step method to determine smoke concentrations and impacts downwind of the burn unit (Wade and Lunsford 1989; Bidwell, Weir, et al. 2003). This procedure will assist burn managers by helping them keep the impact of smoke on the environment and surrounding area within an acceptable limit. The first step is to determine the category day under which the fire should be conducted. Category day is determined by the ventilation rate, which is the afternoon mixing height in meters multiplied by the transport wind speed in meters per second. The mixing height will determine how high into the atmosphere the smoke will rise, and the transport wind speed will establish how rapidly the smoke will move out of the area. The mixing height and transport wind speed, or the predetermined ventilation rate, can be obtained from most local National Oceanic and Atmospheric Administration (NOAA) fire weather Web sites (http://www.noaa.gov/fireweather/), national forest Web sites (www.fs.fed.us), or the national fire weather page (http://fire.boi.noaa.gov/).

Once you have determined the ventilation rate, use Table 14-1 to determine which category day is best for your burn unit. No burning should occur on a category I day, but burning is allowable on category II, III, IV, and V days. Exercise extreme care when burning on a category II day; this type of day allows for only marginal smoke dispersion. You should also be very careful when burning on a category V day; these days have the best smoke dispersion but are also typical of days when thunderstorms can develop. It is advisable not to burn on days when there is a probability of thunderstorms developing.

The second step is to determine how far downwind you should look for smoke-sensitive areas, such as airports, highways, communities, recreation areas, federal Class I areas, schools, hospitals, and residences. Select one or more of the four ignition and burn unit size categories: backfire less than 1,000 acres (405 ha); headfire less than 1,000 acres; any burn greater than 1,000 acres; and brush piles or windrows. If you are using an unlisted firing technique, the burn should be considered a headfire if less than 1,000 acres. Use Table 14-2 to find the appropriate category day and minimum number of miles downwind to screen for SSAs. For example, if you are burning an area that is 160 acres (65 ha) with a ring firing technique, you will use the row labeled "Headfire, less than 1,000 acres." If the burn is going to be conducted on a category II day, you will need to screen for SSAs 20 miles (32.2 km) downwind; on a category IV day, you will need to screen only 5 miles (8.1 km) downwind.

The third step is to use a map or aerial photograph on which SSA locations

TABLE 14-1 Determining category day

Category day	*Ventilation rate*	*Burning guidelines*
I	<2,000	No burning!
II	2,000–4,000	No burning until 11:00 A.M. and not before surface inversion has lifted. Fire should be out by 4:00 P.M.
III	4,000–8,000	Daytime burning only after inversion has lifted.
IV	8,000–16,000	Burn anytime. For night burns, use backfires with surface winds greater than 4 mph (1.8 m/sec).
V	>16,000	Unstable and windy. Excellent smoke dispersion, but burn with caution.

Source: From Bidwell, Weir, et al. (2003).

TABLE 14-2 Distance (miles) downwind that should be examined for smoke-sensitive areas

	Category day				
Type of burn	*I*	*II*	*III*	*IV*	*V*
Backfire: less than 1,000 acres	NB*	10	5	2.5	.75
Headfire: less than 1,000 acres	NB*	20	10	5	.75
Fire: greater than 1,000 acres	NB*	20	10	5	.75
Brush piles or windrows	NB*	30	15	8	.75

Source: From Bidwell, Weir, et al. (2003).
* No Burn.

can be identified. The region on the map or photograph should be large enough to view an area over 20 miles downwind from the burn unit. Locate the burn unit on the map, and draw a line through the center of the burn unit, which represents the wind direction to be used at the time of the burn. This line should extend well past the downwind side of the unit. Then draw two other lines at a 30° angle from this centerline, making sure that the 30° lines are on the outside edge of the burn unit boundary on the map (Figure 14-2). If you are burning a large unit and the 30° lines are not set outside the burn unit, your smoke plume

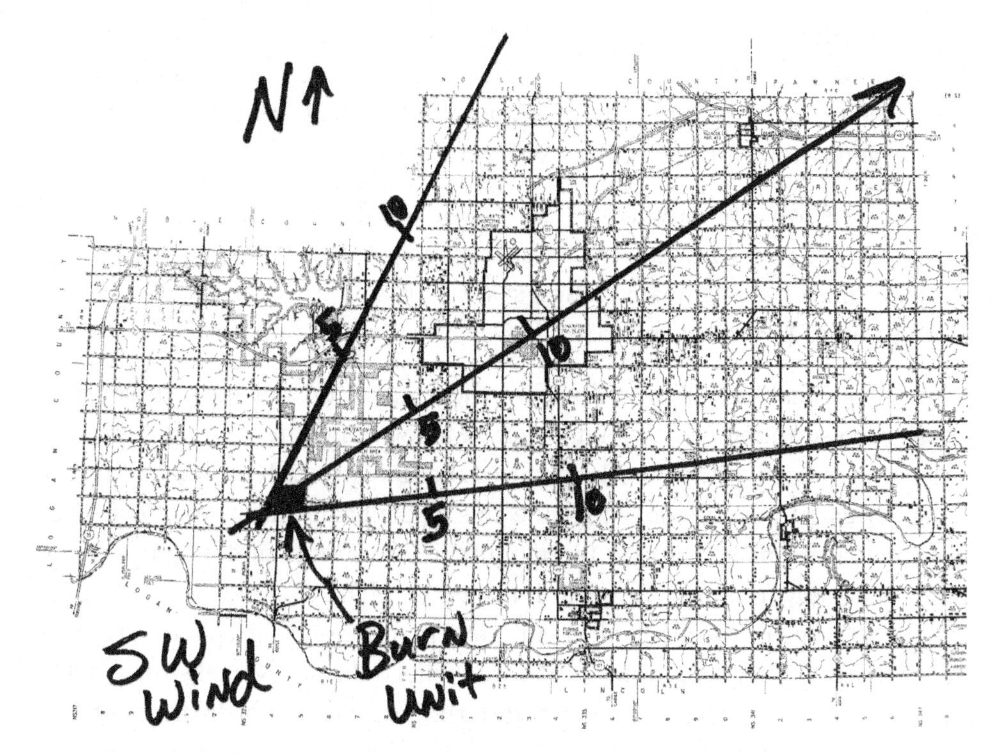

Figure 14-2.
An example of a smoke plume trajectory map. Using a map, draw a line through the center of the burn unit, and then draw two lines 30° from the centerline, which will be the area affected by the smoke plume.

trajectory will be too narrow and possibly not include all of the SSAs that will be affected. Using the scale from the map or aerial photograph, mark the centerline and 30° lines at 5, 10, and 20 miles (8.1, 16.1, and 32.2 km) downwind from the burn unit; this will allow you to determine the category day(s) that can be used. The centerline and 30° lines should be drawn for each wind direction that is planned for burning the unit. If the winds are not very predictable, or are not from a constant direction, use a 45° angle from the centerline to increase the screening area. You can also use a compass to draw circles around the burn unit at the appropriate distances; then screen for SSAs and determine the most suitable wind direction and category day.

Determining if any SSA resides within the smoke plume trajectory area by looking at the smoke plume trajectory map is the fourth step of the smoke screening process. If any other known sources of smoke overlap your smoke plume trajectory area, you should consider increasing the screening distance on the map to account for the effects of the combination of different smoke plumes. You should also consider higher-category days, burning smaller units if multiple fires are going to be conducted, or not burning at all if the smoke impact will be too great for the region. If there is no SSA within the projected area, go ahead and burn as prescribed. If there is an SSA in the projected area, continue with the screening process.

It is also important to recognize critical SSAs that already have air pollution or visibility problems, such as factories. You should identify any potential sulfur dioxide (SO_2) emission areas like smelters, electric power plants, and other factories where coal is used. When SO_2 combines with particulate matter from prescribed fires, significant health hazards are possible. If a critical SSA is located near the projected smoke plume, follow these additional screening recommendations for these areas (Wade and Lunsford 1989):

- 5 miles (8.1 km), any critical SSA within ½ mile (0.8 km)
- 10 miles (16.1 km), any critical SSA within 1 mile (1.6 km)
- 20 miles (32.2 km), any critical SSA within 2 miles (3.2 km)
- 30 miles (48.1 km), any critical SSA within 3 miles (4.8 km)

If any critical SSA is located within these areas, do not burn under your present prescription. You should prescribe a new wind direction, or if the critical SSA is located within the last half of the distance criteria, pick a higher-category day if possible or reduce the size of the burn unit by one-half. Then have the burn finished at least three hours before sunset, and mop up quickly and completely; monitor the burn unit into the night.

The fifth step is to evaluate the results and modify your burn plans as needed. If you identified any SSA in the fourth step that could be adversely impacted by smoke emission from your prescribed burn, postpone the burn and wait for a higher-category day, use a different wind direction, or use a treatment method other than prescribed fire. If there are no SSAs in the smoke plume trajectory, go ahead with the burn, but still consider steps to minimize the amount of smoke produced.

How to Minimize Smoke Problems

You can use many methods to reduce the impact of smoke outside the burn unit. Some of the methods are easy to accomplish, while others may require receiving authorization or permits and can add expense to the burn. The simplest method to reduce smoke problems is to burn smaller units; reduced fuel loads create less smoke. This requires you to conduct more burns, which increases the cost and takes more time, but if it is the only way you can safely conduct prescribed fires, then it is the best method. The following are methods that can be used to reduce smoke problems:

BURN DURING PROPER WEATHER CONDITIONS

Burn when atmospheric conditions are best for rapid smoke dispersal, normally after the morning inversion layer has broken and before the evening inversion layer forms (Nikleva 1972); or when the atmosphere is neutral to unstable, which enhances plume rise and the horizontal and vertical dispersing of smoke (Wade and Lunsford 1989). Conduct nighttime burns only if you have a favorable forecast, because smoke movement and visibility are hard to predict, and nighttime temperature inversions will cause smoke to hold at ground level (Schroeder and Buck 1970).

Burn according to air pollution regulations and not during pollution alerts, stagnant conditions, or ozone alert days because smoke will not disperse rapidly and will instead stay near ground level (Wade and Lunsford 1989). Determine minimum parameters for smoke dispersal by using minimum surface and upper-level wind speeds, desired wind direction, minimum mixing height, category day, and dispersion index (EPA 1998).

Carry out burns during seasons of the year with meteorological conditions that allow for proper smoke dispersion (Nevada Environmental Protection Division 1999). Smoke emissions can be reduced by 50% by burning in one season compared to another (Sandburg and Dost 1990).

Anticipate and minimize down-drainage smoke movement (Wade and Lunsford 1989), especially in mountainous areas where downslope winds prevail at night (Schroeder and Buck 1970). Burn only after evaluating smoke dispersion with a computer model or smoke plume trajectory plot.

BURN WHEN FUEL CONDITIONS ARE MOST ACCEPTABLE

Burn with proper fuel moisture conditions. This can be accomplished by selecting the correct combination of fuel moisture and fuel-size class that needs to be removed to meet the burn objectives. For removal of fine fuels, burn when the relative humidity is low enough for these fuels to burn and when the larger fuels are too wet to ignite.

Use test fires prior to burning to confirm fire and smoke behavior before igniting the entire unit (Wade and Lunsford 1989). This is accomplished by igniting a small area inside the burn unit that can be easily contained and extinguished and then observing how well the smoke lifts and disperses. If conditions are not favorable, extinguish the test burn and wait for better atmospheric conditions.

Estimate the amount of smoke the fuels will produce. This is sometimes difficult to determine and comes with experience. Areas that have not been burned in years will create greater amounts of smoke than frequently burned areas.

Figure emission rates; fuels with high moisture content, high concentration of oils, or large fuel particle size will have higher rates of smoke emissions. Exercise care when burning in areas with organic soils, also known as peat or muck soils; during dry conditions these soil types can ignite and burn for days, weeks, or even months.

REDUCE FUELS TO MANAGE SMOKE

Use periodic maintenance-type prescribed burns that follow historic natural fire return intervals (Nevada Environmental Protection Division 1999). Consolidate nonmerchandisable material in commercial forestry areas, have timber sales of multiple products, use chemical or mechanical treatments, and

Figure 14-3.
Be careful when conducting burns near roads; take all precautions to keep the smoke off the road. When burning near roads, have an emergency plan in case the smoke does reach the roadway.

allow firewood cutting (Nevada Environmental Protection Division 1999). Utilize single or multispecies grazing on rangelands to reduce fine fuels, or use haying practices. Exercise care when using certain mechanical treatments because they can increase the amount of fuel and volatility of fuels within a burn unit.

REDUCE THE IMPACT OF SMOKE ON PEOPLE

Notify all people who could possibly be affected downwind, such as nearby residents, adjacent landowners, fire departments, and local fire control offices. This is common courtesy and common sense, and in some states is also the law. Inform smoke-sensitive people how to avoid smoke exposure. Provide clean-air facilities for sensitive people, and relocate smoke-sensitive people until the risk is over (EPA 1998). Mop up along roads as soon as possible, and pay special attention to roads that are in areas where smoke can travel downslope or down drainage (Wade and Lunsford 1989) (Figure 14-3). Use appropriate signage to inform the public about areas where smoke will impact them, such as highways, secondary roads, trails, and campgrounds (Nevada Environmental Protection Division 1999).

Initiate public education or public relations activities prior to conducting burns. Based on weather conditions and number of burns within a given airshed, use a notification process that requires authorization or approval from an entity before conducting burns, which is already required in certain states (EPA 1998).

IGNITION TECHNIQUES FOR SMOKE MANAGEMENT

Consider burning by using backfires to reduce the amount of smoke produced. Backfires cause higher amounts of fuel to be consumed in the flaming rather than the smoldering stage of combustion (NWCG 1985). Use mass-ignition techniques like ring firing or headfires to create greater amounts of heat, which will create more lift for the smoke. Use mass-ignition devices such as the helitorch, DAID, or terra torch to increase fuel consumption per unit of time and thus limit the duration of smoke and increase convection (Nevada Environmental Protection Division 1999).

POSTBURN MOP-UP TO REDUCE NUISANCE SMOKE

Outline what actions you will take after the fire is over to reduce residual smoke, like prompt or rapid mop-up, mop-up of certain fuels, and complete or 100% mop-up of smoking fuels. If residual smoke problems from logs, snags, or stumps could be a problem, take steps to keep them from burning. If they do ignite, extinguish them quickly. If postburn smoke could be a problem, be sure to monitor the burn unit and have personnel in place to suppress any fuels that begin to smolder.

The fire boss has the responsibility to manage the smoke on each fire. The incorporation of one or more of these smoke impact reduction methods can lessen problems, both current and future, on most prescribed fires. Remember that even when the smoke leaves the burn unit, it is still the burn crew's smoke, and they should do everything possible to reduce the impact on people outside the burn unit.

CHAPTER 15 Postburn Mop-up

Every thing that may abide the fire, ye shall make it go through the fire, and it shall be clean: nevertheless it shall be purified with the water of separation: and all that abideth not the fire ye shall make go through the water.

—Numbers 31:23

The main objective of postburn mop-up is to prevent the residual fire from igniting outside the firebreak; the secondary purpose is to reduce or eliminate postburn smoke problems. Most of the larger spotfires I have encountered on prescribed burns are caused by improper mop-up after the burn. These fires can restart anywhere from a few hours after the burn to a day or two later. Because of the risk of spotfires and postburn smoke problems, mop-up is one of the most crucial phases of prescribed burning and should not be taken lightly. There is nothing more aggravating than having to come back to a burn unit and put out an escaped fire because the fire was not mopped up properly. It also looks bad for the fire boss or prescribed fire program, because many times the crew is already gone when the fire does escape and someone else has to report it.

Mop-up should begin with the ignition phase and continue until the fire boss declares that the fire is out. The initial phase starts with personnel monitoring the perimeter during ignition. This involves either walking or driving the perimeter looking for problem areas. These potential problems could include burning logs, snags, brush piles, unburned fuels, cow chips, or smoldering mulch piles. If any such problem areas are encountered, they should be monitored or dealt with promptly. If you have enough personnel on the burn, assign a couple of people to follow the ignition crew and begin mop-up along the edges.

Preliminary mop-up should begin along the edge of the burn unit and at the hottest spots or areas that can cause the most problems along the edge. Mop-up should then progressively work inward until an area has been secured at a safe or designated distance along the edge of the entire burn unit.

The distance that should be mopped up back from the edge will vary. Some factors that help determine mop-up distance are agency policy, fuel type, fuel load adjacent to burn unit, current weather conditions, and forecasted weather conditions. In most grass fuels, mopping up only 5 to 10 feet (1.5 to 3.0 m) inside the perimeter may be sufficient because the main residual fire problems associated with grass fuels are mulch piles and cow chips. In burn units with shrubs and trees, any brush piles, downed woody debris, or standing snags may require mopping up 20 to 200 feet (6.1 to 60.1 m) back inside the burn unit. Each burn unit is different and will require its own specific mop-up plan.

In some areas where residual smoke is a major problem, cleaning up areas emitting smoke is an important concern. This may involve wetting down a few logs or cutting down a snag. But in areas where any smoke is a problem, you may need to complete a 100% mop-up following a prescribed burn. A 100% mop-up means that nothing will be left burning, smoldering, or smoking inside the entire burn unit.

For example, while in Florida, I assisted with a burn at a local state park. The park policy was for 100% mop-up following all burns; interstate highways ran along two sides of the park, and homes were built right next to the boundary fence. Most burn units in the park were small, averaging only 10 acres (4 ha) each. The unit we burned was 8 acres (3.24 ha), but on that small piece of ground we used 6,000 gallons (22,713 L) of water to mop up after the burn was completed. It took us thirty minutes to burn the unit and three hours to mop up. I figured out quickly why the burn units were so small and that 100% mop-up was not enjoyable but necessary.

Weather is an important factor on mop-up operations. Always plan ahead by checking the weather forecast for the day of the burn, as well as for that evening, that night, and the day after the burn. You can then make the proper mop-up decisions concerning the distances that smoldering material should be cleared from the edge and whether or not personnel will be required to stay overnight or even return the next day. Pay attention to temperatures, relative humidity, and dew point, as well as wind speed and direction.

Wind speed can be a two-edged sword when related to mop-up. Light winds can cause fewer safety concerns about blowing embers; but higher wind speeds will allow most smoldering fuels to burn up faster. Also, in many regions winds die down at night, resulting in slower combustion rates, so fuels do not completely burn down at night and will continue to burn into the next day. The fuels can flare up in the peak heating periods during daylight hours, possibly igniting spotfires. However, many of the fuels will be extinguished due to the light winds and increase in relative humidity.

Lack of wind at night can also cause smoke problems. The light wind allows smoke to accumulate in the burn unit and around adjacent areas. A well-prepared smoke management plan may have worked during the burn, but post-burn smoke and weather conditions also have to be considered. You need to think about what will happen when the wind drops off at night and allows the smoke to accumulate in the areas you had protected during the day.

Determining the forecasted wind direction for the next one to three days following the burn is also important. This will help you determine which side of the burn unit to mop up thoroughly, and which side can be left to burn, thus reducing the amount of work the crew will have to perform. For example, if your burn unit has several large brush piles located along the north firebreak, these piles can burn for several days and be an area of potential fire escape. If the forecasted wind direction for the next three days is from the north, you will have little cause for concern. This northerly wind will blow the embers back into the blackened burn unit. But what if the forecast is for the wind to swing around to the south and become gusty? In this case you should mop up the piles and monitor them until they are burned down.

It is also very important to keep abreast of the forecasted temperature and dew point for several days following the burn because temperature affects the heating and drying of fuels. If the air temperatures are expected to be warm, then care should be taken to mop up thoroughly; if temperatures are predicted to be cooler, mop-up may not need to be so intensive. Many times dew point temperatures will impact mop-up in a positive way. Air temperatures reaching dew point at night will either drastically reduce the burning times of many fuels or put the fires out completely. In areas where dew points are often reached at night, mop-up can be stopped at dark and started the next day or may not even be needed the following day. Many times the fuels absorb enough moisture to limit combustion.

Relative humidity and fuel moisture will also influence the amount of work needed following a prescribed burn; drier conditions mean more work mopping up. Again, keep track of the forecasted relative humidity, and check for any frontal passages following the burn. Fronts can increase relative humidity and put out smoldering fuels, or they can dry out the atmosphere and cause fuels to burn for longer periods with greater intensity. The 10-hour, 100-hour, and 1,000-hour fuel moistures will have great impact on the combustion of fuels and residual smoke. If conditions are dry enough for any of these fuel moisture classes to ignite, mop-up may be a long and tedious task. Therefore, if a burn unit has large amounts of certain fuel moisture classes that will cause problems, it may be best to wait until these fuels are too moist to ignite. But if the objective of your prescribed fire is for a reduction in these fuel moisture classes, be prepared for a fairly large amount of postburn work and possible residual smoke problems.

Mop-up Techniques

When you mop up, one of the main rules to follow is to permit fuels to burn up completely if they will do so promptly and safely. Most of the time a limb, log, or mulch pile can either be left alone or pulled into the unit a short distance and allowed to burn out. This will save water, time, and personnel for more important tasks. So many times people want to spray water on everything that is glowing or smoking. In doing so, they are lulled into a false sense of security. More often than not, the wet, smoldering fuel will dry out and rekindle in half an hour or so when no one is around. So make sure crew members understand when to use water and when to let material burn (Figure 15-1). For example, if the burning or smoldering debris does not cause a safety threat or can be moved farther inside the burn unit, save your water for items that cannot be moved or that have a greater risk for causing an escape.

If snags are burning next to the line and embers are blowing across the line,

Figure 15-1.
Smoldering mulch piles can be a problem during mop-up. Many people think that just spraying water on them will put out the fire. Within half an hour these piles will be burning again. The best way to safely handle them is to let them burn out on their own or drag them farther back into the blackened burn unit.

the snag should be felled, foamed, or soaked down repeatedly, or a person with tools and a radio should be left to monitor the site. Crews may also have to fell snags outside the burn unit to reduce spotting (Figure 15-2). I have observed numerous dry snags catch fire outside a burn unit because of blowing firebrands, but no other fuels in the area ignited. During mop-up in areas with dead trees, be sure to watch out for snags burning outside the burn unit. Snags may also have to be cut down to eliminate safety hazards to personnel. Limbs or whole trees that have burned though can fall without warning, injuring or even killing crew members. Be sure to mark areas with paint or flagging tape, or set up lookouts when personnel are working in areas with burning snags nearby or overhead.

When felling snags, you should make crew safety your top priority. If mechanical means such as a bulldozer is available, use it first. Make sure the driver is protected from falling debris and embers before pushing the snag over. When using a chainsaw to fell snags, cut on a side without limbs if possible and have a person watching out for the sawyer. The person cutting, and anyone nearby, should have on full protective gear. Many times we have a lookout holding on to the back and belt of the person cutting; this person is watching for falling debris and can pull the sawyer back if trouble arises. Care should be exercised when cutting trees that have glowing embers; the air flow from the chainsaw will cause them to flame up rapidly, which could injure the sawyer or damage the saw.

During mop-up it may be necessary to cut lower limbs from fuels to reduce

Figure 15-2.
Embers can travel long distances from the top of a tall snag. Because mopping-up large snags can take a lot of time and water, many times it is best to push the snag into the black or fell the snag.

or eliminate ladder fuels. Many times ladder fuels do not burn during the main fire but become dried out and can ignite later. This could carry the fire into the crown of nearby trees and cause spotting across the fireline.

If brush or trees pose a threat inside or outside the burn unit, remove them before the burn or during mop-up to reduce fire hazards. Stumps near the edge of the unit can be a problem because they cannot be moved very easily. If a stump is burning, it can be dug out and removed or mud packed. Mud packing is a very effective mop-up technique for burning stumps, logs, or snags. It is also a useful technique when a pumper truck cannot get near the site to soak the smoldering material down. Mud packing entails making a slurry of mud and smearing it on the burning portion of the stump or log. This will cover any blowing embers and effectively smother the area to put the fire out. It uses very little water and works very well.

You may also need to remove the duff from under trees, especially if the trees have been dried out by the fire. During one particularly dry prescribed fire season, we had numerous eastern redcedar trees crown out up to three days after the initial fire. The duff under the trees would smolder and then catch fire, creating enough heat and flame to crown out the already scorched cedar tree. If this occurred next to the fireline, it could cause an escape and more problems.

When dealing with logs, cut or break them into pieces that are easily handled. If they are small enough or you have enough personnel, you can drag them a safe distance into the burn unit. If available, an ATV and a chain can be used to pull logs a safer distance into the burn unit. If piles are burning or smoldering near the edge, it is best to scatter the logs within the burn unit to reduce the amount of heat being created by the pile. Use a bulldozer to push them farther into the burn unit; otherwise, drag them individually and scatter them a safe distance inside the burn unit. On slopes it may be essential to reorient the logs that are lying parallel to the edge lengthwise to the slope so they will not roll out of the burn unit. If they cannot be moved, you can create trenches below the logs to stop them from rolling.

In burn units where cattle are present, you may need to kick, rake, or use a leaf blower to move the cow chips back from the edge of the firebreak. Cow chips can smolder for hours and have been the cause of numerous escapes when not properly mopped up.

If any areas near the firebreak contain unburned fuels, be sure to burn these areas out to avoid future problems. A fire could spread quickly across the unburned areas and send embers outside the burn unit. The largest escape I have had was on such a fire. The day after we burned the unit the weather turned warm and dry, so we were back at the unit mopping up. We located an area adjacent to the east firebreak that contained unburned grass fuels and cedar trees. The area had rekindled, but we could not get a pumper unit to the site, so we were working with backpack pumps and hand tools. The southwesterly wind had strong gusts and carried the fire into several large eastern redcedar

Figure 15-3.
During mop-up do not allow crew members to bunch up; make sure they spread out and look for problem areas.

trees, which crowned and spotted over the firebreak. We worked four hours and chased the fire for over a mile before finally containing it. This shows the necessity of removing any potential escape hazards before the burn if possible.

During mop-up, make sure that crew members are spread out around the burn unit perimeter and not bunched up visiting with each other (Figure 15-3). When personnel are bunched up, very little work gets done, and more important, no one is watching the perimeter for possible escapes. Make sure people are moving around and watching for problem areas.

It is important to continue to patrol the firelines until the area is declared safe. This may require several days and nights of personnel and equipment patrolling the area. The main thing to remember is that if you properly mop up, the fire should not go anywhere.

Detection of Hot Materials

During mop-up many potential problem areas are not easy to detect. In order to find problems, you often have to purposely look for them. The best way to find them is to use your senses. Look for smoke, heat waves, or areas of white ash. Feel for areas of heat on the ground or on burned logs. You must take your gloves off and hold the back of your hand above the suspected area to determine if any heat is coming off the area. As you patrol an area, listen closely for crackling material or hissing water vapor escaping from fuels. Last, and the most difficult, use your nose to detect any smoke. This usually can be accomplished several hours after the burn when most of the smoke has dissipated and only the fresh smoke lingers.

If you are not sure if an area is hot, you can use water to assist with finding any hot spots. Using a pumper unit, backpack pump, or canteen, place some water on the suspected area and listen for steam hissing. You should also look for steam rising from the site. Handheld infrared thermometers can be used for detecting hot areas without touching them. These instruments are pointed at the suspected area to take a temperature reading. If the reading is greater than the surrounding area, the spot is hot and needs to be cooled down.

Mop-up Tools

You can use the following tools to help mop up an area. Remember to use the right piece of equipment for the task at hand, and be sure not to sacrifice safety for convenience.

- *Pumper units:* Extinguish, soak, wet, or cool down fuels; patrol fire lines; haul equipment; haul personnel around the unit
- *ATV/Utility vehicles:* Patrol fire lines, haul equipment, haul personnel around the unit
- *Backpack pumps:* Extinguish, soak, wet, or cool down fuels; as a water source for places where pumper units cannot go; spread water out around perimeter of burn unit; monitor hot spots
- *Rakes/McLeods:* Pull debris back inside burn unit, cut down small shrubs and brush, drag limbs and logs, scrape burning areas from logs and snags, mix water into soil to cool down hot spots
- *Brooms:* Sweep debris back inside burn unit
- *Shovels:* Dig up soil to cover stumps, logs, or other burning material; scrape burning areas from logs and snags; mix water into soil to cool down hot spots
- *Chainsaws:* Fell snags, cut logs into manageable size, clear shrubs or brush from under other trees
- *Leaf blowers:* Blow debris back into burn unit along the edge of the firebreak; clear out leaves and duff from under shrubs, brush, and trees
- *Drip torches/Ignition devices:* Ignite material so it will burn out rapidly; burn out unburned fuels near the fireline

Mop-up Safety

During mop-up all personnel should continue to wear their PPE because this phase of prescribed burning can be just as hazardous as any of the others (Figure 15-4). When personnel pick up debris to move it back into the unit, there is a good chance that they may get burned. You should make sure that everyone has on at least gloves and a long-sleeved shirt when working. If working near power equipment, crew members should also wear hearing and eye protection. If the burn is conducted in a wooded area, be sure that everyone is wearing a hard hat because falling limbs, snags, and embers can cause severe injury.

An excellent illustration of this occurred on a burn that we had conducted

Figure 15-4.
Make sure that everyone has on PPE when working after the burn because the potential for injury is still there.

and were mopping up. A dead tree had fallen and was resting on its branches next to the firebreak. Several crew members decided to roll it into the black. They started rolling it; the top half broke off and fell directly on a crew member's head. Even though the tree fell only a few feet, the force knocked the person down; fortunately, he had on a hard hat and was not seriously hurt. The most interesting part of this mishap was that it happened in front of a television news crew doing a story on prescribed burning.

Again, exercising caution when working around and with snags cannot be stressed enough. Extreme care should be exercised when felling trees. When working in areas with steep slopes, be extremely careful of rolling debris; this includes not only burning logs but also rocks and boulders that may become dislodged and roll down onto unsuspecting crew members.

Both the people driving vehicles and those working along the fireline should be watching out for each other. Make sure drivers slow down and do not drive into areas of dense smoke. Make sure that everyone continues to use all of the equipment in a safe and proper manner.

It is important to monitor weather conditions even after the burn is over. If changes occur in temperature, relative humidity, wind speed, or direction, let everyone know and take the appropriate measures to keep the fire contained.

The main thing to remember is not to get too comfortable. We all think that when the fire is over, there is nothing left to worry about. This could not be further from the truth. Never let your guard down; continue to monitor the fire and the weather. Make sure to move personnel and equipment to where they are most needed. Do not waste any time or resources. You will have time to relax once the fire is completely out and all mop-up has been finished.

CHAPTER 16 Prescribed Fire Associations

Ask, and it shall be given you; seek, and ye shall find; knock, and it shall be opened unto you. For every one that asketh receiveth; and he that seeketh findeth; and to him that knocketh it shall be opened.

—Matthew 7:7–8

When you ask people why they do not burn, they will usually give you one of four reasons (McNeill 2003). The number-one reason is liability. Liability is, and should be, a concern for all people involved with prescribed fire, but it should not be a brick wall. Second is their lack of training for or experience with conducting a burn. Third is not having enough people to help conduct a prescribed burn. And the fourth reason is that they do not have enough equipment for setting and controlling a fire.

Instead of giving in to these four excuses, you should ask yourself, "How do I overcome these obstacles?" First, the issue of liability can be addressed with insurance. Most farm and ranch liability policies cover damages that might occur from an escaped fire. (Check with your insurance company to find out what coverage you have.) You can then attend workshops on conducting prescribed fires to gain insight into how to conduct a burn. Cooperative extension agencies, NRCS, or other agencies run workshops in many states. Second, to address the need for labor, simply hire extra personnel; you should be able to find people who are willing to work. The same goes for equipment required to conduct the burn; purchase what is needed. There are numerous equipment vendors out there, so you should be able to find someone to supply you with adequate equipment to meet your specific needs.

Although these answers give you a place to start, they still do not deal directly with the problem. Insurance is fine, but it does not help most people feel more confident about conducting a burn. You probably are still concerned with learning how to better manage your risk when burning. You can attend numerous workshops, but this may not give you the hands-on experience of actually being involved with and conducting prescribed fires (Figure 16-1). You will probably still have some questions, such as, "What activities can help me learn about fire behavior and the important techniques that help reduce the risk of escaped fires?" Also, hiring labor is costly: "Where can I hire people who have prescribed fire experience?" Equipment is also costly, and maybe you have only 160 or 320 acres (64.75 or 129.5 ha) that need to be burned. If you buy equipment and figure it into the cost of burning, then burning is not economically effective for the small landowner.

Many times these reasons are just excuses that can be overcome by asking for help, seeking knowledge, and knocking on your neighbor's door. In fact, a way to overcome all these issues is currently being used: prescribed burning associations. Associations are groups of ranchers/landowners/land managers

Figure 16-1. Prescribed burning associations provide training and hands-on learning about using prescribed fire. Technical assistance is important and can be provided by universities or state and federal agencies.

pooling their resources, sharing their labor, and safely conducting prescribed burns.

These prescribed fire associations deal directly with the four reasons most people will not burn. You still will need to have your own insurance, but you can manage your risk by addressing and improving on the other three reasons why you might be hesitant to burn. You can continue to attend workshops or trainings, but you also will attend actual fires. This helps you gain experience and confidence with prescribed fire. You do not have to hire labor because you now have the situation of neighbor helping neighbor. Also, both you and your neighbor are now experienced in prescribed burning, which helps further reduce the risks. This same group of people then can pool their equipment to save on cost and the worry of where to store the equipment. One person may have a drip torch, another a slip-on pump unit, another a four-wheeler, and another a tractor and disk for preparing firebreaks. This also helps reduce the risk of burning by having more equipment on hand.

How do you get an association started in your community? First, assemble a group of interested citizens. Make sure you involve key members of the community such as landowners, lessees, USDA-NRCS employees, county extension services, other federal or state agencies, and local fire departments. Second, you need a leader. For the association to really work, it has to be a grassroots program with local people in charge. There can and should be technical assistance and guidance from agency and or county extension personnel, but the association will not prosper unless someone from the community steps forward to lead and encourage others. Remember that, as with any organization, you only get out of it what you put into it.

Goals for Rx Fire Associations

The association should then set some goals and have a specific task in mind. As with any plan or organization, there need to be goals and objectives to help guide the way. The four goals listed here are adapted from the Edwards Plateau Prescribed Burning Association, Inc. (EPPBA) in Sonora, Texas, and will help give you an idea of where to start: share equipment; share labor; train the membership; and foster good relations between neighbors and within the community in regard to the use of prescribed fire.

Other associations I have worked with have adopted these same goals. These goals outline what an association should do—teach and train the landowners/managers how to use fire safely and properly, as well as educate those in the community about the positive aspects of fire. The last part is the most important—to have a viable prescribed fire program, you must gain the support of the people of the community.

Guidelines for Rx Fire Associations

A set of guidelines for associations has also been compiled from the various active associations (Taylor 2005; Weir and Bidwell 2005). These guidelines can be used as listed or can be modified to fit your needs.

- *Elect officers:* President, vice-president, and secretary/treasurer and a board of directors (one or two from each county involved); landowners/lessees only; agency/university personnel should only provide technical assistance
- *Dues:* $25.00/year (used to buy equipment)
- *Fire training school (annual):* Safety, equipment use, techniques
- *Fire plans:* Prepared by landholder with help from agency or extension personnel
- *Liability:* Landholder assumes liability for fire and must show proof of insurance before burn
- *Firelines:* Landholder responsible for preparing lines; lines must be adequate to contain the fire
- *Personnel on burn:* Have a minimum number that must be present on each burn
- *Equipment:* Have an inventory of what is available
- *Burn participation:* Once association established, members must assist with a certain number of burns before their own land is burned
- *Officers/Board of directors:* Approves fire plans and addresses needs from each county

These guidelines provide a basis for how the association will function. They also help describe what is expected from the members and what the members should expect from the association.

An association can also incorporate; this allows the association to gain nonprofit status as a 501(c)(3) group. Nonprofit status allows the association the benefit of not having to pay taxes on income; it also makes dues, donations, gifts, and monetary contributions tax deductible. The association then becomes eligible for grants from governmental agencies and private foundations that donate only to nonprofit corporations. The association also becomes eligible for a bulk-mailing permit. For information about obtaining 501(c)(3) status, obtain the Internal Revenue Service form 1023 or talk to a lawyer.

Several prescribed burning associations have been able to use local rural fire department equipment while conducting prescribed burns. The EPPBA was able to purchase a fire truck from a local fire department. The Sonora Volunteer Fire Department (VFD) sold the fire truck to the Edwards Plateau Soil and Water Conservation District so that it would be available to the members of the burning association. The association charges members mileage to use the truck for the round-trip from Sonora to the burn site (Taylor 2001). The Big Pasture Prescribed Burning Association in Oklahoma uses trucks from a VFD when it burns within certain counties in the association (Bennett 2003). Some VFDS may request a small donation for their services. Joining forces with a VFD can have benefits for both parties. It gives the burn association added equipment and possibly more personnel, while it gives the VFDS some training time, added income, and community service. All the while, both parties are having a positive impact on the community.

One of the greatest risk management benefits of a prescribed fire association is that if a fire does escape, it will probably be onto another member's property. The other members understand fire and the risks associated with using it. They also realize that an escape could happen on their fire, so there are usually not any hard feelings or litigation.

One of the main attributes of a prescribed burning association is its ability for strength in numbers. Neighbors are helping neighbors reduce the risks of a wildfire on their own lands through proper land management, and the associations are reducing the risk of wildfires to the other members of their community. Also, you have group support for a management activity that is sometimes misunderstood and portrayed in a less than desirable light. When members of the community band together with the same goal, safely applying fire to the landscape, many community members will become supportive. They will also enjoy the benefits of prescribed fire in their area. These benefits include reduction of volatile fuels for wildfire protection; enhanced wildlife, livestock, and native plant habitat; better water quality; and reduction of brush species that cause allergy and asthma problems. Another important aspect of an association is that when you burn, you are training future generations how to use fire to

Figure 16-2. Prescribed burning associations help future generations become comfortable with the use of prescribed fire. This allows the landowners and community to become at ease with reintroducing fire into the region.

responsibly manage the land and showing why fire plays an important part in the ecosystem (Figure 16-2).

I have been involved in the formation of fourteen prescribed fire associations. One association had an initial formation meeting and within two weeks had conducted five burns on 1,200 acres (485.62 ha). This association did not put a lot of emphasis on details and just started burning; they figured they would concern themselves with the organizational details after the burn season was over. Another association had great attendance and enthusiasm at the beginning but delayed the details of establishment and training so did not carry out any burning for over a year. This type of slow start can hurt attendance and cause a loss of interest, so plan accordingly. Do not try to start an association in the middle or at the end of the burn season because you need time to get at least some of the details in place before you begin burning.

At this time there are prescribed burning associations located in Oklahoma, Texas, Colorado, Kansas, and one in California that has been in existence since 1956 (Dale 1999; Taylor 2005). These prescribed burning associations give the local landowners and managers the equipment, labor, education, and experience to conduct prescribed burns safely and responsibly. They allow the members to become comfortable with the use of fire and to train future generations in the proper application of prescribed fire. These burn associations also provide wildfire reduction practices, public relations, and educational opportunities to the communities where they are located.

The following Web sites provide more information on prescribed burn associations:

- *Oklahoma Prescribed Fire Council:* http://www.oklahomaprescribedfirecouncil.okstate.edu/
- *Edwards Plateau Prescribed Burning Association, Inc.:* http://www.ranchmanagement.org/eppba/
- *Texas Panhandle Prescribed Burn Association:* http://www.ranches.org/tppba.htm
- *Brush Country Prescribed Burn Association:* www.bcpba.com
- *Red Hills Prescribed Burn Association:* http://www.rhpba.com/

References

Agricultural Air Quality Task Force. 1999. *Air quality policy on agricultural burning: Recommendation from the Agricultural Air Quality Task Force to U.S. Department of Agriculture.* Washington, DC: Agricultural Air Quality Task Force.

Alexander, M. E. 1982. Calculating and interpreting forest fire intensities. *Canadian Journal of Botany* 60:349–57.

American National Standards Institute. 1998. Z308.1-1998. *Minimum requirements for industrial unit-type first-aid kits, appendix A.* Washington, DC: American National Standards Institute.

Anderson, H. E. 1968. Fire spread and flame shape. *Fire Technology* 4:51–58.

Anderson, K. L., E. F. Smith, and C. E. Owensby. 1970. Burning bluestem range. *Journal of Range Management* 23:81–92.

Anderson, M. K. 2005. *Tending the wild: Native American knowledge and the management of California's natural resources.* Los Angeles: University of California Press.

Baedecker, P. A., E. O. Edney, P. J. Morgan, T. C. Simpson, and R. S. Williams. 1991. Effects of acidic deposition on materials. In *Acidic deposition: State of science and technology.* Vol. III, *Terrestrial, materials, health and visibility effects,* ed. P. M. Irving. Washington, DC: The U.S. National Acidic Precipitation Assessment Program (NAPAP). *Atmospheric Environment* 26:147–58.

Bailey, A. W., and M. L. Anderson. 1980. Fire temperatures in grass, shrub, and aspen forest communities of central Alberta. *Journal of Range Management* 32:37–40.

Beaufait, W. R. 1966. *Prescribed fire planning in the intermountain West.* USDA Forest Service Res. Paper INT-26. Ogden, UT: Intermediate Forest and Range Experimental Station.

Bennett, A. C. 2003. Burn baby burn! *The Oklahoma Cowman* 42:26–27.

Betchley, C., J. Q. Koenig, G. van Belle, H. Checkoway, and T. Reinhardt. 1997. Pulmonary function and respiratory symptoms in forest firefighters. *American Journal of Industrial Medicine* 31:503–9.

Bidwell, T. G., and J. F. Stritzke. 1989. *Eastern redcedar and its control.* Fact Sheet F-2850. Stillwater: Oklahoma Cooperative Extension Service, Oklahoma State University.

Bidwell, T. G., D. M. Engle, J. R. Weir, R. E. Masters, and J. D. Carlson. 2006. *Fire prescriptions for maintenance and restoration of native plant communities.* Fact Sheet F-2878. Stillwater: Oklahoma Cooperative Extension Service, Oklahoma State University.

Bidwell, T. G., J. R. Weir, and D. M. Engle. 2002. *Eastern redcedar control and management—best management practices to restore Oklahoma's ecosystems.* Fact Sheet F-2876. Stillwater: Oklahoma Cooperative Extension Service, Oklahoma State University.

Bidwell, T. G., J. R. Weir, J. D. Carlson, M. E. Moseley, R. E. Masters, P. McDowell, D. M. Engle, S. D. Fuhlendorf, J. Waymire, and S. Conrady. 2003. *Using prescribed fire in Oklahoma.* Circular E-927. Stillwater: Oklahoma Cooperative Extension Service, Oklahoma State University.

Bidwell, T. G., S. Fuhlendorf, B. Gillen, S. Harmon, R. Horton, R. Manes, R. Rodgers, S. Sherrod, and D. Wolfe. 2003. *Ecology and management of the Lesser Prairie-Chicken in Oklahoma.* Circular E-970. Stillwater: Oklahoma Cooperative Extension Service, Oklahoma State University.

Biswell, H. H., H. R. Kallander, R. Komerak, R. J. Vogl, and H. Weaver. 1973. *Ponderosa fire management: A task force evaluation of controlled burning in ponderosa pine forests in central Arizona.* Misc. Pub. No. 2. Tallahassee, FL: Tall Timbers Research Station.

Bitgood, S., and D. Patterson. 1987. Principles of exhibit design. *Visitor Behavior* 2 (1): 4.

Blair, B. K., J. C. Sparks, and J. Franklin. 2004. Effective application of prescribed burning. In *Brush management past, present, future,* ed. W. T. Hamilton, A. McGinty, D. N. Ueckert, C. W. Hanselka, and M. R. Lee. College Station: Texas A&M University Press.

Boyd, C. S., and T. G. Bidwell. 2001. Influence of prescribed fire on Lesser Prairie-Chicken habitat in shinnery oak communities in western Oklahoma. *Wildlife Society Bulletin* 29:938–47.

Brenner, J., and D. Wade. 1992. Florida's 1990 Prescribed Burning Act: Protection for responsible burners. *Journal of Forestry* 90:27–30.

———. 2003. Florida's revised prescribed fire law: Protection for responsible burners. In *Proceedings of fire conference 2000: The first national congress on fire ecology, prevention, and management,* ed. K. E. M. Galley, R. C. Klinger, and N. G. Sugihara, 132–36. Misc. Pub. No. 13. Tallahassee, FL: Tall Timbers Research Station.

Breyfogle, S., and S. A. Ferguson. 1996. *User assessment of smoke-dispersion models for wildland biomass burning.* Gen. Tech. Rep. PNW-GTR-379. Portland, OR: U.S. Department of Agriculture, Forest Service, Pacific Northwest Research Station.

Britton, C. M., and H. A. Wright. 1971. Correlation of weather and fuel variables to mesquite damage by fire. *Journal of Range Management* 23:136–41.

Brown, A. A., and K. P. Davis. 1973. *Forest fire: Control and use.* New York: McGraw-Hill.

Buck, G. 1935. Les tiques a Madagascar. *Revue Agricole et Sucriere de l'Ile Maurice* 84:196–209.

Bunting, S. C., and H. A. Wright. 1974. Ignition capabilities of nonflaming firebrands. *Journal of Forestry* 72:646–49.

Byram, G. M. 1959. Combustion of forest fuels. In *Forest fire: Control and use,* ed. K. P. Davis, 61–89. New York: McGraw-Hill.

Canon, S. K., P. J. Urness, and N. V. DeByle. 1987. Habitat selection, foraging behavior, and dietary nutrition of elk in burned aspen forest. *Journal of Range Management* 40:433–38.

Carpenter, J. W., H. E. Jordan, and B. C. Morrison. 1973. Neurologic disease in wapiti naturally infected with meningeal worm. *Journal of Wildlife Diseases* 9:148–53.

Catchpole, E. A., N. J. de Mestre, and A. M. Gill. 1982. Intensity of fire at its perimeter. *Australian Forest Research* 12:47–54.

Chapman, R. N., D. M. Engle, R. E. Masters, and D. M. Leslie Jr. 2004. Tree invasion constrains the influence of herbaceous structure in grassland bird habitats. *Ecoscience* 11:55–63.

Cheney, P., and A. Sullivan. 1997. *Grassfires: Fuels, weather and fire behavior.* Collingwood, Australia: CSIRO Publishing.

Coppedge, B. R., D. M. Engle, R. E. Masters, and M. S. Gregory. 2001. Avian response to landscape change in fragmented southern Great Plains grasslands. *Ecological Applications* 11:47–59.

Cram, D. S., T. T. Baker, J. Boren, and C. Edminster. 2008. Classification of wildland fire effects in silviculturally treated vs. untreated forest stands of New Mexico and Arizona. In *Proceedings of the 2002 fire conference: Managing fire and fuels in the remaining wildlands and open spaces of the southwestern United States,* technical coordinator Marcia G. Narog, 332–33. Gen. Tech. Rep. PSW-GTR-189. Albany, CA: U.S. Department of Agriculture National Forest Service. Pacific Southwest Research Station.

Dale, D. 1999. Management by fire. *Beef* (August):28–30.

Daniel, T. C. 1990. Social/political obstacles and opportunities in prescribed fire management. In *Proceedings: Effects of fire management of southwestern natural resources,* ed. J. S. Krammes. November 15–17, 1988. GTR-RM-191. Tucson, AZ: USDA Forest Service, Rocky Mountain Forest and Range Experiment Station, Fort Collins, CO.

DeBano, L. F., D. G. Neary, and P. F. Ffolliot. 1998. *Fire's effects on ecosystems.* New York: Wiley.

DeByle, N. V., P. J. Urness, and D. L. Blank. 1989. *Forage quality in burned and unburned aspen communities.* Res. Paper INT-404. Ogden, UT: U.S. Department of Agriculture, Forest Service, Intermountain Research Station.

Dubé, D. 1977. Prescribed burning in Jasper National Park. *Forest Report* 5:4–5. Canadian Forest Service, Northern Forest Research Centre, Edmonton, Alberta.

Engle, D. M., and J. F. Stritzke. 1992a. Enhancing control of eastern redcedar through individual plant ignition following prescribed burning. *Journal of Range Management* 45:493–95.

———. 1992b. Herbage production around eastern redcedar trees. In *1983–1991 range research highlights,* 13–14. Circular E-905. Stillwater: Oklahoma Cooperative Extension Service, Oklahoma State University.

Engle, D. M., and J. R. Weir. 2000. Grassland fire effects on corroded barbed wire. *Journal of Range Management* 53:611–13.

Engle, D. M., J. R. Weir, D. L. Gay, and B. P. Dugan. 1998. Grassland fire effects on barbed wire. *Journal of Range Management* 51:621–24.

Fahnestock, G. R., and R. C. Hare. 1964. Heating of tree trunks in surface fires. *Journal of Forestry* 62:799–805.

Fenner, R. L., R. K. Arnold, and C. C. Buck. 1955. *Area ignition for brush burning.* USDA Forest Service Tech. Paper No. 10. Berkeley, CA: Pacific Southwest Forest and Range Experimental Station.

Firecon, Inc. N.d. *Gel fuel torch operation manual.* Ontario, OR: Firecontrol, Inc.

Fosberg, M. A. 1970. Drying rates of heartwood below fiber saturation. *Forest Science* 16:57–63.

Franklin, S. E. 1993. Chaparral management techniques: An environmental perspective. *Fremontia* 21:21-24.

Fraser, C. M., J. A. Bergeron, A. Mays, and S. E. Aiello, eds. 1991. *The Merck veterinary manual.* 7th ed. Rahway, NJ: Merck.

Froelich, R. C., C. S. Hodges Jr., and S. S. Sackett. 1978. Prescribed burning reduces severity of annosus root rot in the South. *Forestry Science* 24:93–100.

Fuhlendorf, S. D., and D. M. Engle. 2001. Restoring heterogeneity on rangelands: Ecosystem management based on evolutionary grazing patterns. *BioScience* 51:625–32.

Gartner, F. R., and W. W. Thompson. 1972. Fire in the Black Hills forest-grass ecotone. *Proceedings of the Tall Timbers Fire Ecology Conference* 12:37–68.

Green, L. R. 1977. *Fuel breaks and other fuel modifications for wildland fire control.* USDA Handbook Number 499. Washington, DC: U.S. Department of Agriculture.

Haines, T. K., and R. L. Busby. 2001. Prescribed burning in the South: Trends, purpose, and barriers. *Southern Journal of Applied Forestry* 25:149–53.

Hardy, C. C., R. D. Ottmar, J. L. Peterson, J. E. Core, and P. Seamon, eds. 2001. *Smoke management guide for prescribed and wildland fire: 2001 edition.* PMS 420-2. Boise, ID: National Wildfire Coordinating Group.

Harrison, H. 1995. *A user's guide to "PUFFX": A dispersion model for smoke management in complex terrain.* Mercer Island, WA: WYNDSoft.

Heady, H. F. 1960. *Range management in East Africa.* Nairobi: Government of Kenya.

Heirman, A. L., and H. A. Wright. 1973. Fire in medium fuels of West Texas. *Journal of Range Management* 26:331–35.

Hogue, L. D. 2000. *Operations Division policy guidance memorandum no. 30.* Tulsa, OK: U.S. Tulsa District Office, Army Corps of Engineers.

Hummel, J., and J. Rafsnider. 1995. TSARS plus smoke production and dispersion model user's guide. Preliminary Draft 7.95. National Biological Service and the Interior Fire Coordination Committee. Unpublished report. On file with Environmental Science and Technology Center (ESTC), 2401 Research Blvd., Suite 205, Fort Collins, CO 80526.

Inter-Agency Prescribed Fire Course. 1999. Public relations. In *Basic prescribed fire training.* Tallahassee: Florida Division of Forestry.

Jacobson, S. K. 1999. *Communication skills for conservation professionals.* Washington, DC: Island Press.

Johnson, E. A., and K. Miyanishi. 1995. The need for consideration of fire behavior and effects in prescribed burning. *Restoration Ecology* 3:271–78.

Johnston, F. H., A. M. Kavanagh, D. M. J. S. Bowman, and R. K. Scott. 2002. Exposure to bushfire smoke and asthma: An ecological study. *Medical Journal of Australia* 176:535–38.

Knopf, F. L. 1986. Changing landscapes and the cosmopolitism of the eastern Colorado avifauna. *Wildlife Society Bulletin* 14:132–42.

Koehler, J. T. 1992. Prescribed burning—a wildfire prevention tool? *Fire Management Notes* 53:9–13.

Launchbaugh, J. L., and C. E. Owensby. 1978. *Kansas rangelands: Their management based on a half century of research.* Bulletin No. 622. Manhattan: Kansas Agriculture Experimental Station

Lavdas, L. G. 1996. *Program VSMOKE—user's manual.* Gen. Tech. Rep. SRS-6. Asheville, NC: U.S. Department of Agriculture, Forest Service, Southeastern Research Station.

Lewis, H. T. 1973. Patterns of Indian burning in California: Ecology and ethnohistory. In *Ballena Anthropology Paper 1,* ed. L. J. Bean. Ramona, CA: Ballena Press.

Lewis, H. T., and T. A. Ferguson. 1999. Yards, corridors, and mosaics: How to burn boreal forest. In *Indians, Fire and the Land in the Pacific Northwest,* ed. R. Boyd, 164–84. Corvallis: Oregon State University Press.

Lindenmuth, A. W., Jr., and G. M. Byram. 1948. Headfires are cooler near the ground than backfires. *Fire Control Notes* 9:8–9.

Lindenmuth, A. W., and J. R. Davis. 1973. *Predicting fire spread in Arizona oak chaparral.* Research Paper RM-101. Fort Collins, CO: Rocky Mountain Forest and Range Experiment Station.

Lowe vs. Jones et al. N.d. Case No. CJ95-345. Oklahoma District Court records available at http://www.odcr.com/ (accessed June 8, 2006).

MacKenzie, T., and B. Fortier. 2004. Prescribed fire cuts insurance premiums. *USFWS Fish and Wildlife News* (Fall issue):4-5.

Martin, G. G. 1988. Fuels treatment assessment: 1985 fire season in Region 8. *Fire Management Notes* 49:21–24.

Martin, R. E. 1990. Goals, methods, and elements of prescribed burning. In *Natural and prescribed fire in Pacific Northwest forests,* ed. J. D. Walstad, S. R. Radosevich, and D. V. Sandberg, 55–66. Corvallis: Oregon State University Press.

Martin, R. E., R. W. Cooper, A. B. Crow, J. A. Cuming, and C. B. Phillips. 1977. Report of task force on prescribed burning. *Journal of Forestry* 75:297–301.

Martinez, D. 1998. Wilderness with or without you. *Earth First!* 18:1–13.

Masters, R. E., K. Hitch, W. J. Platt, and J. A. Cox. 2005. Fire—the missing ingredient for natural regeneration and management of southern pines. *Proceedings of the Joint Conference, Society of American Foresters and Canadian Institute of Forestry,* October 2–6, 2004, Edmonton, Alberta, Canada. CD-ROM.

Masters, R., and J. Waymire. 2000. *The effects of timber harvest and fire frequency on wildlife and wildlife habitat in the Ouachita Mountains.* Stillwater, OK: Oklahoma Agricultural Experiment Station.

McMahon, C. K. 1999. Forest fires and smoke—impacts on air quality and human health in the USA. In *Proceedings of the TAPPI International Environmental Conference,* April 18–21, 2:443–543. Nashville, TN: TAPPI PRESS.

McNeill, A. 2003. Ring of fire. *The Cattleman* 89:10–14.

McPherson, G. R., G. A. Rasmussen, H. A. Wright, and C. M. Britton. 1986. *Getting started in prescribed burning.* Management Note 9. Lubbock: College of Agricultural Sciences, Texas Tech University.

McWhiney, G. 1988. *Cracker culture: Celtic ways in the old south.* Tuscaloosa: University of Alabama Press.

Miller, W. E. 1978. Use of prescribed burning in seed production areas to control red pine cone beetle. *Environmental Entomology* 7:698–702.

Mitchell, R. G. 1990. Effects of prescribed fire on insect pests. In *Natural and prescribed fire in Pacific Northwest forests,* ed. J. D. Walstad, S. R. Radosevich, and D. V. Sandberg, 111–16. Corvallis: Oregon State University Press.

Mobley, H. E. 1976. Smoke management—what is it? In *Southern forest fire laboratory personnel. Southern smoke management guidebook.* Gen. Tech. Rep. SE-10. Asheville, NC: U.S. Department of Agriculture, Forest Service, Southeastern Forest Experiment Station.

Montana/Idaho State Airshed Group. 2005. http://www.smokemu.org/ (accessed February 8, 2005).

Mutch, R. W. 1970. Wildland fires and ecosystems—a hypothesis. *Ecology* 61:1046–51.

National Park Service. 1983. *Interpretive planning handbook.* Washington, DC: Government Printing Office.

———. 1999. *Wildland fire reference manual #18.* PDF file available at http://www.nps.gov/fire/fire/fir_wil_pla_reference18.html (accessed July 17, 2005).

National Wildfire Coordinating Group. 1985. *Smoke management guide.* PNW-420-2. NFES 1279. Boise, ID: National Interagency Fire Center, National Interagency Coordinating Group, Prescribed Fire and Fire Effects Working Team.

———. 1986. *Firefighters guide.* NFES 1571. Boise, ID: National Interagency Fire Center, National Interagency Coordinating Group, Fire Equipment Working Team.

———. 1986. *Prescribed fire plan guide.* NFES 1839. Boise, ID: National Interagency Fire Center, National Interagency Coordinating Group, Prescribed Fire and Fire Effects Working Team.

———. 1991. *Firing methods and procedures: s-234.* NFES 2029. Boise, ID: National Interagency Fire Center.

———. 1992. *Foam vs. fire primer.* NFES 2270. Boise, ID: National Interagency Fire Center.

———. 1994a. *Fire effects guide.* NFES 2394. Boise, ID: National Interagency Fire Center.

———. 1994b. *Water handling equipment guide.* NFES 1275. Boise, ID: National Interagency Fire Center, National Interagency Coordinating Group, Prescribed Fire and Fire Effects Working Team.

———. 1996. *Wildland fire suppression tactics reference guide.* NFES 1256. Boise, ID: National Interagency Fire Center.

Nevada Environmental Protection Division. 1999. *Nevada smoke management program.* Carson City, NV: Dept. of Conservation and Natural Resources.

Nikelva, S. 1972. The air pollution potential of slash burning in southwestern British Columbia. *Forestry Chronicles* 48:187–89.

Owensby, C. E., and E. F. Smith. 1979. Fertilizing and burning Flint Hills bluestem. *Journal of Range Management* 32:254–58.

Owsley, F. L. 1945. *Plain folk of the south.* Baton Rouge: Louisiana State University Press.

Packard, S., and C. F. Mutel, eds. 1997. *The tallgrass restoration handbook: For prairies, savannas, and woodlands.* Washington, DC: Island Press.

Partanen, T. 1993. Formaldehyde exposure and respiratory cancer—a meta-analysis of the epidemiological evidence. *Scandinavian Journal of Work, Environment, and Health* 19:8–15.

Philpot, C. W. 1969a. *The effect of reduced extractive content on the burning rate of aspen leaves.* Ogden, UT: USDA–Forest Service Research Note INT-RN-92.

———. 1969b. *Seasonal changes in heat content and ether extractive content in chamise.* Ogden, UT: USDA–Forest Service Research Paper INT-RP-61.

Pyne, S. J. 1982. *Fire in America: A cultural history of wildland and rural fire.* Seattle, WA: University of Washington Press.

Pyne, S. J. 1984. *Introduction to wildland fire: Fire management in the United States.* New York: Wiley.

Pyne, S. J., P. L. Anderson, and R. D. Laven. 1996. *Introduction to wildland fire.* New York: Wiley.

Raskevitz, R. F., A. A. Kocan, and J. H. Shaw. 1991. Gastropod availability and habitat utilization by wapiti and white-tailed deer sympatric on range enzootic for meningeal worm. *Journal of Wildlife Diseases* 27:92–101.

Reinhardt, T. E., and R. D. Ottmar. 1997a. Employee exposure review. In *Health hazards of smoke: Recommendations of the April 1997 Consensus Conference,* ed. B. Sharkey, 29–40. Tech. Rep. 9751-2836-MTDC. Missoula, MT: USDA–Forest Service, Missoula Technology and Development Center.

———.1997b. *Smoke exposure among wildland firefighters: A review and discussion of current literature.* Gen. Tech. Rep. PNW-GTR-373. Portland, OR: USDA–Forest Service, Pacific Northwest Research Station.

———. 2000. *Smoke exposures at western wildfires.* Research Paper PNW-RP-525. Portland, OR: USDA–Forest Service, Pacific Northwest Research Station. Roberts, K. W., D. M. Engle, and J. R. Weir. 1999. Weather constraints to scheduling prescribed burns. *Rangelands* 21:6–7.

Roscommon Equipment Center. 1976. *Jeep tanker handbook.* Project Number 4. Roscommon, MI: Northeast Forest Fire Supervisors, in cooperation with Michigan's Forest Fire Experiment Station.

———. 1988. *M-561 "Gamma Goat": An analysis of the vehicle for wildfire use.* Project Number 53-A. Roscommon, MI: Northeast Forest Fire Supervisors, in cooperation with Michigan's Forest Fire Experiment Station.

———. 1990a. *Fire control use of all terrain vehicles.* Project Number 55. Roscommon, MI: Northeast Forest Fire Supervisors, in cooperation with Michigan's Forest Fire Experiment Station.

———. 1990b. *Hard cab and slip-on tank designs for the Gamma Goat (M-561).* Project Number 53B. Roscommon, MI: Northeast Forest Fire Supervisors, in cooperation with Michigan's Forest Fire Experiment Station.

———. 1990c. *Vehicles for fire management.* Project Number 48. Roscommon, MI: Northeast Forest Fire Supervisors, in cooperation with Michigan's Forest Fire Experiment Station.

Rothermel, R. C. 1983. *How to predict the spread and intensity of forest and range fires.* USDA–Forest Service Gen. Tech. Rep. INT-143. Ogden, UT: Intermountain Forest and Range Experiment Station.

Ryan, P. W., and C. K. McMahon. 1976. Some chemical characteristics of emissions from forest fires. Paper No. 76-2.3. In *Proceedings of the 69th annual meeting of the Air Pollution Control Association,* Pittsburgh, PA. Portland, OR: Air Pollution Control Association.

Sackett, S. S. 1975. Scheduling prescribed burns for hazard reduction in the Southeast. *Journal of Forestry* 73:143–47.

Sammons, J. H., and R. L. Coleman. 1974. Firefighters' occupational exposure to carbon monoxide. *Journal of Occupational Medicine* 33:1163–67.

Sandburg, D. V., and F. N. Dost. 1990. Effects of prescribed fire on air quality and human health. In *Natural and prescribed fire in Pacific Northwest forests,* ed. J. D. Walstad, S. R. Radosevich, and D. V. Sandburg, 191–218. Corvallis: Oregon State University Press.

Sandburg, D. V., R. D. Ottmar, J. L. Peterson, and J. Core. 2002. *Wildland fire on ecosystems: Effects of fire on air.* Gen. Tech. Rep. RMRS-GTR-42-vol. 5. Ogden, UT: U.S. Department of Agriculture, Forest Service, Rocky Mountain Research Station.

Schroeder, M. J., and C. C. Buck. 1970. *Fire weather: A guide for application of meteorological information to forest fire control operations.* Agricultural Handbook 360. Washington, DC: USDA–Forest Service.

Scire, J., D. G. Strimaitis, R. J. Yamartino, and Z. Xiaomong. 1995. *A user's guide for CALPUFF dispersion model.* Doc. 1321-2. Concord, MA: Sigma Research/Earth Tech.

Scrifes, C. J., and W. T. Hamilton. 1993. *Prescribed burning for brushland management: The South Texas example.* College Station: Texas A&M University Press.

Seamon, P., ed. 2004. *Fire management manual.* The Nature Conservancy. PDF file available at http://www.tnc.fire.org/manual (accessed July 16, 2005).

Seitel, F. P. 2001. *The practice of public relations.* Upper Saddle River, NJ: Prentice Hall.

Sestak, M. L., and A. R. Riebau. 1988. *SASEM simple approach smoke estimation model.* Tech. Note 382. Denver, CO: U.S. Department of Interior, Bureau of Land Management.

Sestak, M. L., W. E. Marlatt, and A. R. Riebau. 1989. VALBOX: Ventilated valley box model. Unpublished report on file with Michael Sestak, U.S. Department of Interior, Bureau of Land Management, and Colorado State University, Environmental Science and Technology Center, 2401 Research Blvd, Suite 205, Fort Collins, CO 80526.

Sharkey, B. J. 1991. *Health hazards of smoke.* Prepared by Missoula Technology and Development Center, USDA–Forest Service. Winter Issue.

———. 1997a. *Fitness and work capacity.* 2d ed. Tech. Rep. 9751-2814-MTDC. Missoula, MT: USDA–Forest Service, Missoula Technology and Development Center.

———. 1997b. Health effects of exposure. In *Health hazards of smoke: Recommendations of the April 1997 Consensus Conference,* ed. B. Sharkey, 41–47. Tech. Rep. 9751-2836-MTDC. Missoula, MT: USDA–Forest Service, Missoula Technology and Development Center.

———, ed. 1997c. *Health hazards of smoke: Recommendations of the April 1997 Consensus Conference.* Tech. Rep. 9751-2836-MTDC. Missoula, MT: U.S. Department of Agriculture, Forest Service, Missoula Technology and Development Center.

———. 1999. Heat stress. In *Wildland firefighter health and safety: Recommendations of the Consensus Conference, April 1999,* ed. B. Sharkey, 33–38. Tech. Rep. 9951-2841-MTDC. Missoula, MT: USDA–Forest Service, Missoula Technology and Development Center.

Simmons, G. A., J. Mahar, M. K. Kennedy, and J. Ball. 1977. Preliminary test of prescribed burning for control of maple leaf cutter (Lepidoptera: Incurvariidae). *The Great Lakes Entomologist* 10:209–10.

Smith, E. F., and C. E. Owensby. 1972. Effects of fire on true prairie grasslands. *Proceedings of the Tall Timbers Fire Ecology Conference* 12:9–22.

Smith, J. K., ed. 2000. *Wildland fire in ecosystem: Effects of fire on fauna.* Gen. Tech. Rep. RMRS-GTR-42-vol. 1. Ogden, UT: U.S. Dept. of Agriculture Forest Service, Rocky Mountain Research Station.

Stanton, R. 1995. Managing liability exposures associated with prescribed fires. *Natural Areas Journal* 15:347–52.

Stanturf, J. A., D. D. Wade, T. A. Thomas, A. Waldrop, D. K. Kennard, and G. L. Achtemeier. 2002. Background paper: Fire in southern forest landscapes. In *Southern forest resource assessment,* eds. D. M. Wear and J. Greis, 607–30.Gen. Tech. Rep. SRS-53. Asheville, NC: U.S. Department of Agriculture, Forest Service, Southern Research Station.

Taylor, C. A. 2001. *Organizing ranchers for prescribed burning: The Edwards Plateau Prescribed Burning Association, Inc.* TR-01-2. Sonora, TX: Texas A&M University.

———. 2005. Prescribed burning cooperatives: Empowering and equipping ranchers to manage rangelands. *Rangelands* 27:18–23.

Taylor, J. G., and R. W. Mutch. 1985. Fire in wilderness: Public knowledge, acceptance, and perceptions. In *U.S. Forest Service General Technical Report INT-GTR-212,* 49–59. Ogden, UT: USDA–Forest Service, Intermountain Research Station.

Thies, W. G. 1990. Effects of prescribed fire on diseases of conifers. In *Natural and prescribed fires in Pacific Northwest forests,* ed. J. D. Walstad, S. R. Radosevich, and D. V. Sandberg, 117–21. Corvallis: Oregon State University Press.

Truesdell, P. S. 1969. Postulates of the prescribed burning program of the Bureau of Indian Affairs. *Proceedings of the Tall Timbers Fire Ecology Conference* 9:235–40.

USDI–Bureau of Land Management. 1984. *Firefighters common task manual.* NFES 1384. Boise, ID: USDI–Bureau of Land Management.

———. 1998. *Interim air quality policy on wildland and prescribed fires.* Washington, DC: U.S. Environmental Protection Agency.

Vale, T. R. 2002. The pre-European landscape of the United States: Pristine or Humanized. In *Fire, Native Peoples, and the Natural Landscape,* ed. T. R. Vale. Washington, DC: Island Press.

Van Amburg, G. L., J. A. Swaby, and R. H. Pemble. 1981. Response of arthropods to a spring burn of a tallgrass prairie in northwestern Minnesota. *Ohio Biological Survey Biological Notes* 15:240–43.

Van der Smissen, B. 1990. *Legal liability and risk management for public and private entities.* New York: Anderson Publishing.

Wade, D. D., and J. D. Lunsford. 1989. *A guide for prescribed fire in southern forests.* USDA–Forest Service Tech. Pub. R8-TP-11. Atlanta, GA: USDA–Forest Service.

Ward, D. E., N. Rothman, and P. Strickland. 1989. *The effects of forest fire smoke on firefighters: A comprehensive study plan.* Prepared for Congressional Committee on Appropriations for Title II-Related Agencies. USDA-Forest Service and the National Wildfire Coordination Group. Ogden, UT: USDA-Forest Service, Intermountain Research Station.

Weaver, H. 1967. Fire and its relationship to ponderosa pine. *Proceedings of the Tall Timbers Fire Ecology Conference* 7:127–49.

Weir, J. R. 2000. A Gator™ for Rx fire. *Rangelands* 22:28–29.

———. 2007. Using relative humidity to predict spotfire probability on prescribed burns. In *Proceedings: Shrubland dynamics—fire and water,* August 10–12, 2004, Lubbock, TX, comp. R. E. Sosebee, D. B. Wester, C. M. Britton, E. D. McArthur, and S. G. Kitchen. Proceedings RMRS-P-47. Fort Collins, CO: U.S. Department of Agriculture, Forest Service, Rocky Mountain Research Station.

Weir, J. R., and T. G. Bidwell. 2005. *Prescribed fire associations.* Fact Sheet F-2880. Stillwater: Oklahoma Cooperative Extension Service, Oklahoma State University.

Weldon, L. A. C. 1996. Dealing with public concerns in restoring fire to the forest. In *The use of fire in forest restoration,* ed. C. C. Hardy and S. F. Arno. Gen. Tech. Rep. INT-GTR-341. Ogden, UT: USDA–Forest Service, Intermountain Research Station.

Whelan, R. J. 1995. *The ecology of fire.* Cambridge: Cambridge University Press.

Williams, M. 1992. *Americans and their forests.* New York: Cambridge University Press.

Wink, R. L., and H. A. Wright. 1973. Effects of fire on an Ashe juniper community. *Journal of Range Management* 26:326–29.

Wolfolk, J. S., E. F. Smith, R. R. Schalles, B. E. Brent, L. H. Harvers, and C. E. Owensby. 1975. Effects of nitrogen fertilization and late-spring burning of bluestem range on diet and performance of steers. *Journal of Range Management* 28:190–93.

Wright, H. A. 1974. Range burning. *Journal of Range Management* 27:5–11.

Wright, H. A., and A. W. Bailey. 1982. *Fire ecology: United States and southern Canada.* New York: Wiley.

Yoder, J. 2002. *Prescribed fire: Liability, regulation, and endogenous risk.* Long Beach, CA: American Agricultural Economics Association Meetings.

Yoder, J., M. Tilley, D. Engle, and S. Fuhlendorf. 2003. Economics and prescribed fire law in the United States. *Review of Agricultural Economics* 25 (1): 218–33.

Index

Agricultural Air Quality Task Force, 159, 183
Air masses, 43–46
Front, three types
Cold, 44–45;
Stationary, 45;
Warm, 45;
High-pressure systems, 44;
Hurricanes, 45;
Low-pressure systems, 44
Aldehydes, 82, 157
Acrolein, 157;
Formaldehyde, 82, 157
Anderson, H. E., 55, 183
Anderson, K. L., 4, 183
Anderson, M. K., 1, 183
Anderson, M. L., 52, 183
Anderson, P. L., 54–57, 186
Anthropogenic fires, 1–2
European settlers, 1;
History of, 1–2;
Native Americans, 1
Aromatic oils, 11
Artemisia filifolia (*see* sand sagebrush)
Ashe juniper, 58
ATV torch, 129

Baedecker, P. A., 157, 183
Bailey, A. W., 9, 52, 65, 68, 90, 93, 99, 136, 147, 149–151, 153, 157, 183, 188
Beaufait, W. R., 153, 183
Benefits of prescribed fire, 6, 8, 11, 23, 31, 33, 35, 180
Accessible browsing, 8;
Distribution of livestock, 6;
Elimination of young trees, 9;
External parasite control; ticks, 6, 8;
Increased production of some oak species, 8;
Reduction of competition, 10;
Reduction of stand density, 8;
Thinned dense stands of trees, 9, 141
Bennett, A. C., 180, 183
Betchley, C., 82, 183
Bible, 1
Deuteronomy, 51;
Exodus, 12, 22;
Ezekiel, 62;
Genesis, 1, 123;
Hebrews, 1;
Isaiah, 78, 168;
James, 88;
Joel, 41;
Matthew, 27, 177;
Numbers, 168;
Revelation, 156;
I Kings, 136;
II Thessalonians, 71
Bidwell, T. G., vii, 2–4, 7, 17, 47, 61, 63, 65, 72, 99, 160–162, 179, 183, 188
Biomass, 10, 183
Bitgood, S., 34, 183
Blackjack oak, 8
Blair, B. K., 58–59, 183
Blank, D.L., 8, 194
Booming grounds, 7
Boyd, C. S., 4, 183
Brenner, J., 15–17, 71, 183
Britton, C. M., 68, 183, 185, 188
Broomweed, 89, 138
Brown, A. A., 54, 184
Browse, 8
Brush control, 7, 65, 76
Brush Country Prescribed Burn Association, 182
Brush species, 4, 6, 11, 148, 180
Dogwood (*Cornus spp.*), 4;
Mesquite (*Prosopis spp.*), 4;
Sumac (*Rhus spp.*), 4;
Oak (*Quercus spp.*), 4;
Other junipers, 4, 101
Buck, C. C., 44, 60, 139, 153, 164–165, 184, 187
Buck, G., 6, 184
Bunting, S. C., 67, 136, 184
Burning conditions, 4, 12, 19
Higher relative humidity, 4;
Lower temperature, 4
Burning season, 9
Rotation, 9
Burn unit, 71–72, 74–75, 77, 83–85, 88–91, 93–96, 99–100, 102–105, 111–112, 114–115, 117, 122–127, 129–131, 133–145, 147–154, 159, 161–171, 173–175
Firebreaks, 88–91, 93–96, 99–100, 102–105;
Fire plans, 71–72, 74–75, 77;
Ignition devices and techniques, 122–127, 129–131, 133–145, 147–154;
Postburn mop-up, 168–171, 173–175;
Size, 6, 149, 159, 161, 164;
Smoke management, 161–167
Busby, R. L., 27, 184
Byram, G. M., 52, 55, 184–185

CAFS (*see* compressed-air-foam systems)
Canon, S. K., 8, 184
Carpenter, J. W., 8, 184
Cervus elaphas (*see* elk)
Chapman, R.N., 7, 184
Cheney, P., 53–54, 58, 137, 146, 184
Cladium jamaicensis (*see* sawgrass)
Clean Air Act, 157–158
Amendments, 158;
History of, 157–158
Clothing (*see* protective clothing)
Coleman, R. L., 82, 187
Combustion, 41, 44, 51, 146, 156–157, 160, 167, 169–170
Four stages of, 51, 156
Flaming stage, 51, 156;
Glowing stage, 51, 156;
Pre-ignition stage, 51, 156;
Smoldering stage, 51, 156
Compressed-air-foam systems, 119
Cooperative extension service, 2, 6
Coppedge, B. R., 7, 184
Cornus spp. (*see* brush species)

DAID (*see* Delayed Aerial Ignition Device)
Dale, D., 181, 184
Daniel, T. C., 30, 184
Davis, J. R., 68, 185
Davis, K. P., 54, 184
Dead fuels, 58, 60
DeBano, L. F., 57–58, 156–157, 184
DeByle, N. V., 8, 184
Delayed Aerial Ignition Device, 134–135, 167
Department of Defense, 107

Defense Reutilization Marketing Service, 107
Diet, wildlife species (*see* wildlife species diet)
Distribution of livestock benefits, 6
Animal performance improvement, 2, 6;
Carrying capacity increase, 2, 6;
Forage quantity and quality improvement, 6
Division of Forestry, 15, 73, 160
Documentation of prescribed fires, 11, 62
Importance of, 11
Dost, F. N., 156, 165, 187
Drip torch, 80–81, 86, 106, 109, 123–127, 129, 131, 138, 140, 144, 175, 178
Benefits of, 124;
How it works, 125;
Operation of, 124–126
DRMS (*see* Department of Defense: Defense Reutilization and Marketing Service)
Dube, D., 93, 184

Eastern redcedar, 2–4, 6–7, 11, 72, 127, 142, 149–150, 173
Edwards Plateau Prescribed Burning Association, Inc., 179–180, 182, 187
Engle, D. M., vii, 2–3, 7, 48–49, 97, 127, 183–184, 186, 188
EPA (*see* U.S. Environmental Protection Agency)
EPPBA (*see* Edwards Plateau Prescribed Burning Association, Inc.)

Fahnestock, G. R., 52, 147, 184
Federal agencies, 19, 21, 108, 178
Federal prescribed fire program, 21
Ferguson, S.A., 160, 183
Ferguson, T. A., 1, 185
Ffolliot, P. F., 55, 57–58, 156–157, 184
Fine fuel loads, 4, 58–59, 65
Fire, causes of, 1–2
Lightning, 1;
Human-caused, 1
Firebreaks, 65, 73, 84–85, 88–93, 95–105, 111–112, 138, 142–143, 147–150, 153, 178
Benefits, 88;
How to build, 88;
Problems, 90–91;
Types of, 92–105;
Daylighting, 102–103;
Disk/tilled/plowed, 99;
Dozed/scraped lines, 98–99;
Fireline plow, 99–101;
Hand lines, 101;
Mowed/wet lines, 93–95;
Natural breaks, 101–102;
No lines, 92;
Other, 103–105;
Roads, 96–98;
Trails, 95–96;
Uses, 88–89;
Width of, 91–92
Fire-derived ecosystems, 8
Fire frequency, 4–5, 8, 31
Fireguards (*see* firebreaks)
Fire intensity, 41, 54–56, 89, 93, 95, 138, 143, 146, 152
Byram's intensity equation, 55;
Three methods of measurement, 55;
Fireline intensity, 55;
Heat per unit, 55;
Reaction intensity, 55
Fire law, 13–14, 17, 21
Contributory negligence, 17–18;
Three types of, 15–17;
Negligent unless proven otherwise, 15;
Not negligent unless proven negligent, 15–17;
Strict or unlimited liability, 15
Firelines (*see* firebreaks)
Fire management program, 8
Control methods, 8
Fire plans, 26, 71–72, 77, 179
Fire boss and crew actions, 71;
Information to include, 72–76;
Legal purposes, 71;
Range of acceptable weather conditions, 71
Fire, prescribed (*see* prescribed fire)
Fire prescription, 62–64, 70
Meeting goals/objectives, 62;
Organized burn, 62;
Safety, 62
Fire, public image of, 27–29
Three aspects of, 28;
Attitude, 28;
Professionalism, 28;
Proper equipment, 28
Fire return interval, 4, 58, 165
Recommended intervals, 4
Fire suppression, 1–2, 11, 16, 21, 113
Fire triangle, 51
Fuel, 51;
Oxygen, 51;
Heat, 51
Fire-tolerant tree species, 9
Roof sprouting, 9;
Serotinous cones, 9;
Underground rhizomes, 9
Fire weather kit, 109–110
Contents, 109
First aid, 84
Flare guns, 130–131
Flame depth, 54
Flame height, 52–53, 93, 137, 150
Flame length, 10, 42, 52–53, 89, 95, 144
Florida Division of Forestry, 160
Forage, 2–3, 5–6, 8, 72
Loss of forage, 3;
Rate of loss, 3
Forbs, 7, 58, 99
Forest disease problems, 10
Forest management, 9
Forestry, 9, 15, 70, 73, 121, 160, 165
Fortier, B., 11, 185
Franklin, J., 58–59, 183
Franklin, S. E., 10, 184
Fraser, C. M., 8, 184
Froelich, R. C., 10, 184
Front, weather (*see* air masses)
Fuel characteristics, 55–59
Classification, 58;
Components, 57–58;
Crown fuels, 58;
Ground fuels, 58;
Surface fuels, 58;
Continuity, 58;
Load, 56–57;
Fuel architecture, 57;
Moisture, 58
Fuel loading, 3, 51, 54, 57, 152
Fuel moisture samples, 61
Fuels, volatile, 11, 17–18, 75, 136, 149–150, 180
Fuhlendorf, S. D., vii, 7, 183–184, 188
Fusee, 126, 138

Gartner, F. R., 147, 184
Gastropods, 8
Georgia Prescribed Burning Act, 16
Governmental agencies, 19, 106, 180
Green, L. R., 68, 184
Gutirezia dracunculoides (*see* broomweed)

Habitat requirements for animals, 6–7
Habitat specialists, 7
Grassland obligate bird species, 7
Haines, T. K., 27, 184
Hamilton, W. T., 62, 147, 149, 183, 187
Hand flares, 130
Hand tools, 85–86, 88, 93, 111–112, 114, 131, 173
Backpack pump, 113–114;
Broom, 112;
Fire rake/McLeod, 111–112;
Leaf blower, 114;
Shovel, 112–113;
Swatter, 112
Hardy, C. C., 156, 184, 188
Hare, R. C., 52, 147, 184
Harrison, H., 160, 185
Harvest operations, 10
Heady, H. F., 6, 185
Heat stress, 78–81, 108
Heat cramps, 78;
Heat exhaustion, 78;
Heat stroke, 79;
Hypothermia, 81;
Prevention, 79–81;
Smoke, 82
Heirman, A. L., 147, 185
Helitorch, 133–134, 167
Herbaceous fuels, 61, 136
Herptofauna, 9
Lizards, 9;
Snakes, 9;
Tortoises, 9
Heterogeneous landscape, 7
High-intensity fires, 10
Hodges Jr., C. S., 10, 184
Hogue, L. D., 25, 185
Hummel, J., 161, 185

Ignition, 1, 8, 51, 56, 68, 71, 73–75, 81–83, 89, 93–95, 114–115, 120–121, 123–127, 129–142, 144–148, 150–156, 159, 161, 167–168, 175
Devices, 124–135;
Types of, 124–135;
Hazards, 136–141;
Canyons, 138–139;
Heavy fuels, 137;
Incomplete burnout, 137–138;
Irregular shape, 138;
Personnel entrapment, 136;
Slopes, 136–137;
Spotfires, 136;
To personnel, 139–141;
Vegetation lines, 139;
Wind changes, 136;
Within the burn unit, 141–144;
Methods, 144–155;
Backfire, 145–147;
Flank-firing, 150–151;
Headfire, 147–150;
Other, 153–155;
Strip headfire, 151–153;
Spot ignition, 153
Insurance, 11, 17–21, 24, 26, 177–179
Premiums, 11
Requirements, 17
Inter-Agency Prescribed Fire Course, 28, 185

J. ashei (*see* Ashe juniper)
Jacobson, S. K., 27–29, 32, 36–37, 185
Johnston, F. H., 157, 185
Jordan, H. E., 8, 184
Juniperus monosperma (*see* shortleaf pine)
Juniperus pinchotii (*see* redberry juniper)
Juniperus spp. (*see* brush species)
Juniperus virginiana (*see* eastern redcedar)

Knopf, F. L., 7, 185
Kocan, A. A., 8, 186
Koehler, J. T., 11, 185

Land management, vii, 2, 4, 6, 11, 18, 21–22, 32, 62, 180
Land management consultant, 2
Launchbaugh, J. L., 147, 185
Lavdas, L. G., 160, 185
Laven, R. D., 54–57, 186
Lawsuits, 13–14, 19
Defendant, 13, 15;
Lowe vs. Jones et al. Case No. CJ95–34, 13;
Plaintiff, 13, 15, 17
Lesser Prairie-Chicken, 7
Liability, 11, 13–21, 24, 62, 67, 70–71, 73, 141, 177, 179
Liability laws, 14, 16;
Managing liability, 18, 24;
Elimination, 18;
Risk retention, 19;
Risk reduction or risk management, 19;
Transfer of risk, 18
Lighter, 123–124, 130
Lindenmuth, A. W., 52, 68, 185
Livestock production, 2, 4–5, 8, 10, 31
Living fuels, 58
Herbaceous, 58;
Woody, 58
Lewis, H. T., 1, 185
Logging areas, management (*see* management of logging areas)
Logging operation, 10, 103
Low-intensity fires, 11, 112, 114, 138, 150
Lunsford, J. D., 10, 72, 146–147, 150–153, 161, 164–166, 188

MacKenzie, T., 11, 185
Management of logging areas, 10
Management tools of prescribed fire, 6
Maintenance of vegetative structure, 6;
Reclamation, 6;
Wildlife habitat improvement, 6
Marlatt, W. E., 160, 187
Martin, G. G., 11, 185
Martin, R. E., 53, 93, 185
Martinez, D., 1, 185
Masters, R. E., 4–5, 8, 183–185
Matches, 35, 124
McMahon, C. K., 156, 158, 185, 187
McNeill, A., 177, 185
McPherson, G. R., 65, 99, 185
McWhiney, G., 1, 185
Meningeal brain worm, 8
Michigan Department of Natural Resources, 120
Miller, W. E., 10, 185
Mitchell, R. G., 10, 185
Mobley, H. E., 156, 186
Moisture content, 43, 55, 57–58, 60–61, 133, 165
Equilibrium moisture content, 60–61
Montana/Idaho State Airshed Group, 159, 186
Morrison, B. C., 8, 184
Mosaic landscape (*see* heterogeneous landscape)
Mutch, R. W., 30, 40, 57, 186–187
Mutel, C. F., 147, 186

NAAQS (*see* National Ambient Air Quality Standards)
National Ambient Air Quality Standards, 158–159

National Association of Foresters, 120
National Fire Plan, 160
National Fire Protection Association, 108
National Oceanic and Atmospheric Administration, 49–50, 161
National Park Service, 25–26, 35
  Wildland and Prescribed Fire Management Policy, 25;
    Fuels management, 26;
    Monitoring, 26;
    Plans, 26;
    Safety, 26;
    Training, 26
Natural regeneration, 9–10
Natural Resource Conservation Service, 2, 6, 70, 92, 177–178
National Weather Service, 48–50
National Wildfire Coordinating Group, 51–55, 57–58, 60, 72, 86, 119–120, 133–134, 146–147, 153, 156, 167, 185–186
Neary, D. G., 55, 57–58, 156–157, 184
Nevada Environmental Protection Division, 165–167, 186
NFPA (*see* National Fire Protection Association)
NOAA (*see* National Oceanic and Atmospheric Administration)
NWCG (*see* National Wildlife Coordinating Group)
NWS (*see* National Weather Service)
No-burn zone, 11
Nonvolatile fuels, 11
NRCS (*see* Natural Resource Conservation Service)

Oak, 4–5, 8, 31, 89, 93, 103, 127, 135, 171, 173, 175
ODWC (*see* Oklahoma Department of Wildlife Conservation)
Oklahoma Department of Agriculture, Food, and Forestry, 121
  Rural Fire Protection program, 121
Oklahoma Department of Wildlife Conservation, 33
Oklahoma Prescribed Fire Council, 182
One-seeded juniper, 101
Osceola National Forest, 2
Ottmar, R. D., 82, 157, 184, 186–187
Owensby, C. E., 4, 147, 183, 185–188
Owsley, F. E., 1, 186

Packard, S., 147, 186
Parelaphostrongylus (*see* meningeal brain worm)
Partanen, T., 82, 186
Patterson, D., 34, 183
Pemble, R. H., 7, 188
Phenolics, 11
Philpot, C. W., 57, 186
Ping-pong ball system (*see* Delayed Aerial Ignition Device)
Pinus echinata (*see* shortleaf pine)
Plant species, 2–3, 8–10, 63
  Control methods, 8
    Herbicides, 8, 137
Postburn mop-up, 168–176
  How to, 168–170;
  Problem detection, 174;
  Purpose of, 168;
  Safety, 175–176;
  Techniques, 170–174;
  Tools, 175
Prescribed fire, vii-182
  Associations, 177–182;
    Goals, 179;
    Guidelines, 179–182;
    Starting one, 178–179;
    501(c)(3) status, 180;
  Benefits (*see* benefits of prescribed fire);
  Equipment, 106–122;
    (*see also* protective clothing, fire weather kit, hand tools, radio, pumper unit, and vehicle)
  Fire behavior and fuel characteristics, 51–61;
    Types of prescribed fire, 52–53;;
    (*see also* flame length, flame height, flame depth, fuel characteristics, fire intensity, rate of spread, and residual flame time)
  Fire weather, 41–50;
    (*see also* temperature, relative humidity, dew point, air masses, wind, weather conditions, weather information, weather patterns, and weather prescription)
  Guidelines for smoke management, 161–164;
  Ignition devices, 123–135;
    (*see also* ignition)
  Ignition techniques, 136–155;
    (*see also* ignition)
  Law and liability, 13–21;
    (*see also* liability)
  Personal safety, 78–87;
    (*see also* safety, personal)
  Plans, 71–78;
    (*see also* fire plans)
  Policy, 22–26;
    Contact information, 24;
    Crisis, 22–24;
    Emergency procedures, 24;
    Emotion, 22–23;
    Safety requirements, 24;
    Smoke management requirements, 24;
  Prescription, 62–70;
    Rules of prescribed fire, 65–66;
    (*see also* fire prescription)
  Public relations, 27–40;
    Public perception of, 22, 27, 29;
    Publics' problems with fire, 29–30;
    (*see also* public relations regarding fire)
  Smoke management, 156–167;
    (*see also* smoke emissions and smoke management)
  Why conduct prescribed fires, 1–12;
    First prescribed fire, 2;
    Management tool, as a, 6
Propane torch, 126–128, 135
Prosopis spp. (*see* brush species)
Protective clothing, 78, 106–109, 126
  Gloves, 108;
  Goggles, 108;
  Helmet, 108;
  Personal protective equipment (PPE), 106;
  Respirator/Face cover, 109;
  Shroud, 108;
  Wildland fire-type clothing, 106–107;
    Indura FR, 106–107;
    Nomex, 106–107
Public perception (*see* prescribed fire: public perception)
Public relations regarding fire, vii, 23–24, 27, 29, 36–38, 40, 166, 181
  Media opinion, 29;
  Objective view, 29;
  Plan/Campaign, 23, 27, 36–37, 40
Public relations tools, 32–40
  Dealing with the "real" public, 40;
  Interviews, 37;
  Letters, signs and posters, 36;

Literature, pamphlets and fact sheets, 35–36;
One-on-One, 37;
Press releases, 36–37;
Talks, forums and public meetings, 32–34;
Tours, field trips, and demonstrations, 34;
Working with the media, 37–40
Pumper unit, 83, 85, 115, 117–122, 173, 175
Foam, 118–119;
Types of, 116–117
Pyne, S. J., 1, 54–57, 136, 138, 186

Q. havardii (*see* sand shinnery oak)
Q. marlandica (*see* blackjack oak)
Q. stellata (*see* post oak)
Quercus spp. (*see* oak)

Radio, 36, 48, 114–115, 123, 140, 150, 171
Types of, 115
Rafsnider, J., 161, 185
Rangeland Ecology and Management group, 22–23
Raskevitz, R. F., 8, 186
Rate of spread, 41–42, 52, 55–56, 58, 146
Reasons for prescribed fires, 2
Cost, 2;
Habitat enhancement, 2
Reasons to burn, 5
Early green-up, 5;
Improved palatability, 5;
Increased nutrient content of new growth, 5;
Overall increase in diversity of plant community, 5
Removal of old forage growth, 5;
Redberry juniper, 58
Red Hills Prescribed Burn Association, 182
Reinhardt, T. E., 82, 157, 183, 186
Relative humidity, 4, 41–45, 60–61, 63–65, 67–72, 78, 109–110, 149, 165, 169–170, 176
Local factors that affect relative humidity, 43–44
Residual flame time, 54, 128
Respondeat superior, 20–21, 24
Rhus spp. (*see* brush species)
Riebau, A. R., 160, 187
Road flare (*see* fusee)
Roberts, K. W., 48–49, 186
Roscommon Equipment Center, 115–116, 119–120, 186
Rothermel, R. C., 53, 55, 187
Rothman, N., 82, 188
Ryan, P. W., 156, 187

Sackett, S. S., 10, 147, 184, 187
Safety, personal, 78, 82, 106, 125
First-aid kits, 84;
Other concerns, 84–87;
Clothing, 86;
Equipment, 85–86;
Food, 85;
Vehicles, 86–87;
Prescribed fire orders, 83;
Prescribed fire situations that shout 'watch out,' 83
Sammons, J. H., 82, 187
Sandburg, D. V., 156, 165, 187
Sand sagebrush, 89
Sand shinnery oak, 4–5, 31, 89
Sawgrass, 104–105
Saw-palmetto, 89
Schroeder, M. J., 44, 60, 139, 164–165, 187
Scire, J., 161, 187
Scrifes, C. J., 147, 149, 187
Seamon, P., 26, 184, 187
Seasonal conditions, 50
Seitel, F. P., 27, 187
Serenoa repens (*see* saw-palmetto)
Sestak, M. L., 160, 187
Sharkey, B., 78–79, 82, 157, 186–187
Shaw, J. H., 8, 186
Shortleaf pine, 4, 103
Simmons, G. A., 10, 187
Smith, E. F., 4, 183, 186–188
Smith, J. K., 146–147, 187
Smoke dispersion models, 159–161
CALPUFF, 161;
NFSPUFF, 161;
SASEM, 160;
TSARS Plus, 161;
VALBOX, 160;
VSMOKE, 160;
VSMOKE-GIS, 160
Smoke emissions, 10, 156–158, 165
Airborne particles, 156, 160;
Carbon dioxide, 156;
Carbon monoxide, 82, 157;
Obstruction of visibility, 157;
Other components, 157;
Water vapor, 43–44, 51, 156, 174
Smoke management, vii, 22, 24, 72–73, 156–159, 161–167, 169
Conditions for, 164–167;
Acceptable fuel conditions, 165;
Fuel reduction, 165–166;
Ignition techniques, 167;
Postburn mop-up, 167;
Proper weather conditions, 164–165;
Smoke impact reduction, 166;
Programs, 158–159
Smoke-sensitive area, 160–161
SMP (*see* smoke management: programs)
Snags, 8, 167–168, 170–173, 175–176
Sparks, J. C., 58–59, 183
Spokesperson, 40
Spotfires, 32, 46, 58, 61, 65, 67–70, 83, 88–89, 93, 97, 108, 111–114, 118, 120–121, 133, 136, 138–141, 147–149, 168–169
Causes/warnings, 138–141, 147–149;
Prevention, 67–70;
Suppression of, 111–114, 121
SSA (*see* smoke-sensitive area)
Stanton, R., 18, 21, 187
Stanturf, J. A., 2, 187
Strickland, P., 82, 188
Stritzke, J. F., 3–4, 127, 183–184
Sullivan, A., 53–54, 58, 137, 146, 184
Suppression, fire (*see* fire suppression)
Swaby, J. A., 6, 188

Tallgrass Prairie Preserve, 92, 121
TGPP (*see* Tallgrass Prairie Preserve)
Taylor, C. A., 179–181, 187
Taylor, J. G., 30, 40, 187
Techniques, prescribed fires (*see* prescribed fire: techniques)
Temperature (weather), 41–49, 51–53, 55–56, 60–61, 63–65, 67, 71–72, 78–81, 109–111, 125–126, 130–131, 134, 144, 147, 149, 164, 169–170, 175–176
Effect on combustion, 51–53;
Effect on rate of spread, 55;
Local factors affecting temperature, 41;
Planning in regard to, 71–72;
Prescriptions regarding, 63–65, 67
Terra torch, 131, 133, 135, 167

Texas Panhandle Prescribed Burn Association, 182
The Nature Conservancy, 26, 92, 101, 121
Thies, W. G., 10, 187
Thompson, W. W., 147, 184
TNC (*see* The Nature Conservancy)
Tools (*see* hand tools)
Treatment options, 2
  Chemical treatments, 2;
  Mechanical treatments, 2
Typanuchus pallidicinctus (*see* Lesser Prairie Chicken)

United States, 1–2, 8–9, 14, 16, 43, 45, 69–70, 92, 99, 124, 134, 138, 158
Urness, P. J., 8, 184
U.S. Army Corps of Engineers, 25, 33
  Tulsa Division, 25
U.S. Environmental Protection Agency, 157–159, 165–166, 188
U.S. Forest Service, 160
  Airfire Team, 160
Ultra vires, 20–21
USACE (*see* U.S. Army Corps of Engineers)

Vale, T. R., 1, 18, 78, 156, 188
Van Amburg, G. L., 6, 188
van der Smissen, B., 18–21, 188
Vegetation regions, 31
Vehicles, 86–87, 115–122
  All-terrain vehicle (ATV), 119–123, 127, 129, 135, 173, 175;
  Requirements, 119–120;
  Utility vehicle, 119–122, 175
Vertical obstructions, 7
VFD (*see* volunteer fire department)
Volunteer fire department, 180
Volatile fuels (*see* fuels, volatile)

Wade, D. D., 10, 15–17, 71–72, 146–147, 150–153, 161, 164–166, 183, 187–188
Ward, D. E., 82, 188
Waymire, J., 4, 8, 183, 185
Weather conditions, 24, 26, 32, 44, 47–48, 50, 52, 62–63, 70–73, 76, 84, 106, 109, 117, 136, 141, 144, 146–150, 152–154, 159, 164, 166, 168–169, 176
  Effect on ignition, 136, 141, 144, 146–150, 152–154;
  Effect on postburn mop-up, 169, 176;
  Effect on smoke management, 159, 164, 166, 168;
  Plans regarding, 71–73, 76;
  Prescriptions regarding, 62–63, 70;
  (*See also* air masses, dew point, relative humidity, temperature, weather information, weather patterns, weather prescriptions, and wind)
Weather information, 48–50
  Forecasts, 48–50;
  Sources, 48–50
Weather patterns, 45, 50
Weather prescriptions, 32
Weaver, H., 9, 183, 188
Weir, J. R., vii, 2–3, 48–49, 68–69, 72, 97, 99, 120, 136, 147, 160, 162, 179, 183–188
Weldon, L. A., 29–30, 188
Whelan, R. J., 55–56, 188
Wildfires, 2, 9–10, 13, 18, 22, 32, 36, 82, 112, 121, 158, 160, 180, 186
Wildlife habitat, 6, 8, 65, 72, 146, 153
Wildlife species diet, 7
Wildland/urban interface, 10, 12
Wind, 41–49, 52–53, 55–56, 63–68, 71–74, 78, 83, 90–91, 96, 99, 109–111, 119, 124, 127, 136, 138–139, 143–155, 161–166, 169–170, 173, 176
  Effect on fire intensity, 55–56;
  Effect on fire types, 52–53;
  Effect on flame length/height, 53;
  Effect on ignition, 143–155;
  Effect on smoke management, 161–166;
  Local factors affecting wind, 46–47
Wink, R. L., 147, 188
Wolfolk, J. S., 4, 188
Wright, H. A., 9, 65, 67–68, 90, 93, 99, 136, 147, 149–151, 153, 157, 183–185, 188

Xanthocephalum dracunculoides (*see* broomweed)

Yoder, J., 13, 15–16, 18, 188

CPSIA information can be obtained
at www.ICGtesting.com
Printed in the USA
LVOW03s0802311215
468357LV00008B/71/P

9 781603 44134